Lecture Notes in Mathematics 1593

Editors:
A. Dold, Heidelberg
B. Eckmann, Zürich
F. Takens, Groningen

Subseries:
Mathematisches Institut der Universität
und Max-Planck-Institut für Mathematik
Bonn - vol. 21

Advisor:
F. Hirzebruch

Jay Jorgenson & Serge Lang
Dorian Goldfeld

Explicit Formulas
for Regularized Products
and Series

Springer-Verlag

**Berlin Heidelberg New York
London Paris Tokyo
Hong Kong Barcelona
Budapest**

Authors

Jay Jorgenson
Serge Lang
Mathematics Department
Box 208283 Yale Station
10 Hillhouse Ave
New Haven CT 06520-8283, USA

Dorian Goldfeld
Mathematics Department
Columbia Unversity
New York, NY 10027, USA

Mathematics Subject Classification (1991): 11M35, 11M41, 11M99, 30B50, 30D15, 35P99, 35S99, 42A99

Authors Note: there is no MSC number for regularized products, but there should be.

ISBN 3-540-58673-3 Springer-Verlag Berlin Heidelberg New York

CIP-Data applied for

Typesetting: Camera-ready T$_E$X output by the authors
SPIN: 10130182   46/3140-543210 - Printed on acid-free paper

EXPLICIT FORMULAS FOR REGULARIZED PRODUCTS AND SERIES

Jay Jorgenson and Serge Lang

A SPECTRAL INTERPRETATION OF WEIL'S EXPLICIT FORMULA

Dorian Goldfeld

EXPLICIT FORMULAS FOR REGULARIZED PRODUCTS AND SERIES

Jay Jorgenson and Serge Lang

A SPECTRAL INTERPRETATION OF WEIL'S EXPLICIT FORMULA

Dorian Goldfeld

EXPLICIT FORMULAS FOR REGULARIZED PRODUCTS AND SERIES

Jay Jorgenson and Serge Lang

Introduction

Explicit formulas in number theory were originally motivated by the counting of primes, and Ingham's exposition of the classical computations is still a wonderful reference [In 32]. Typical of these formulas is the Riemann-von Mangoldt formula

$$\sum_{p^n \le x} \log p = x - \sum_{\rho} \frac{x^{\rho}}{\rho} - \zeta'_{\mathbf{Q}}/\zeta_{\mathbf{Q}}(0) - \frac{1}{2}\log(1 - x^{-2}).$$

Here the sum on the left is taken over all prime powers, and the sum on the right is taken over the non-trivial zeros of the Riemann zeta function.

Later, Weil [We 52] pointed out that these formulas could be expressed much more generally as stating that the sum of a suitable test function taken over the prime powers is equal to the sum of the Mellin transform of the function taken over the zeros of the zeta function, plus an analytic term "at infinity", viewed as a functional evaluated on the test function.

It is the purpose of these notes to carry through the derivation of the analogous so-called "explicit formulas" for a general zeta function having an Euler sum and functional equation whose fudge factors are of regularized product type. As a result, our general theorem applies to many known examples, some of which are listed in §7 of [JoL 93c]. The general Parseval formula from [JoL 93b] provides an evaluation of the "term at infinity", which we call the Weil functional. Also, as an example of our results, let us note that even in the well-studied case of the Selberg zeta function of a compact Riemann surface, our computations show that one may deal with a larger class of test functions than previously known.

For some time, analogies between classical analytic number theory and spectral theory have been realized. Minakshisundaram-Pleijel defined a zeta function in connection with the Laplacian on an arbitrary Riemannian manifold [MiP 49], and subsequently Selberg defined his zeta function [Se 56]. In [JoL 93a,b] we developed a general theory of regularized products and series applicable equally to the classical analytic number theory and to some of these analogous spectral situations. In particular, we proved the basic properties of the Weil functional at infinity in the context of regularized products and series, with a view to using the functional for the explicit formulas in this general context.

A fundamental class of zeta functions. In [JoL 93c] we defined a fundamental class of functions to which we could apply these properties and carry out analogues of results in analytic number theory. Roughly speaking, the functions Z in our class are those which satisfy the three conditions:

  - there is a functional equation;
  - the logarithm of the function admits a generalized Dirichlet series converging in some half plane (we call this Dirichlet series an Euler sum for Z);
  - the fudge factors in the functional equation are of regularized product type.

The precise definition of our class of functions is recalled in Chapter II, §1. The explicit formula can be formulated and proved for functions in this class. In Chapter II, §1, we discuss the extent to which this class is a much broader class than a certain class defined by Selberg [Se 91]. Furthermore, certain applications require an even broader class of functions to which all the present techniques can be applied. We shall describe the need for such a class in greater detail below.

Just as we did for the analogue of Cramér's theorem proved in [JoL 93c], we emphasize that the explicit formula involves an inductive step which describes a relation between some of the zeros and poles of the fudge factors and some of those of the principal zeta function Z. Such a step can be viewed as a step in the ladder of regularized products, because our generalized Cramér theorem insures that a function Z in our class is also of regularized product type provided the fudge factors are of regularized product type.

If Z is a function in our class, and, for $\mathrm{Re}(s)$ sufficiently large, the expression

$$\log Z(s) = \sum_{\mathbf{q}} \frac{c(\mathbf{q})}{\mathbf{q}^s}$$

is the Euler sum for $\log Z(s)$, with a sequence $\{\mathbf{q}\}$ of real numbers > 1 tending to infinity, and complex coefficients $c(\mathbf{q})$, then such $\mathbf{q}$ play the role of prime powers. However, readers should keep in mind cases when $\mathbf{q}$ does not look at all like a prime power. For example, the general theory applies to the case when $Z(s)$ is a general

Dirichlet polynomial, up to an exponential fudge factor; a precise definition is given in Chapter II, §4. Such polynomials contain as special cases the local factors of more classical zeta functions and L-functions. In examples having to do with Riemannian geometry, $\log \mathbf{q}$ is the Riemannian distance between two points in the universal covering space.

The general version of Cramér's theorem in [JoL 93c] was carried out for the original Cramér's test function $\phi_z(s) = e^{sz}$. One can also view this version as a special case of an explicit formula with more general test functions. This is carried out in Chapter II. In [JoL 93c], §7 we gave a number of examples for our Cramér-type theorem. To these we are adding not only the general Dirichlet polynomials as mentioned above, but also Fujii-type L-functions, obtained from a zeta function by inserting what amounts to a generalized character as coefficient of the Dirichlet series defining the zeta function (see the papers by Fujii listed in the bibliography). In Chapter V we show both how to recover Fujii's theorems for the functions he considered, namely the Riemann zeta function and the Selberg zeta function for $PSL(2, \mathbf{Z})$, as well as an analogous theorem for the general zeta functions in our class, all as corollaries of our Cramér's theorem. Similarly, a result of Venkov, which relates the eigenvalues of the Laplacian relative to $PSL(2, \mathbf{Z})$ to the classical von Mangoldt function, will be generalized to any non-compact finite volume hyperbolic Riemann surface in [JoL 94]. The generalization involves another inductive type argument, using the fact that the fudge factor in the functional equation of the non-compact Selberg zeta function involves the determinant of the scattering matrix, which itself is in our class of functions since it has an Euler sum and a simple functional equation with constant fudge factors. In this case, the Euler sum exists whereas a classical Euler product does not. Thus, the general theory simultaneously contains previous results and gives new ones which were not proved previously by authors using such tools as the Selberg trace formula.

Analytic estimates for the proof. In addition to the Parseval formula of [JoL 93b], the proof of the general explicit formula relies on certain analytic estimates for regularized harmonic series, including the logarithmic derivatives of regularized products in strips. We gave such estimates already in [JoL 93a,b], but we need further such estimates which we present in Chapter I of the present work, using the technique of our generalized Gauss formula. Hard-core analytic estimates having thus been put out of the way, the rest of the work is then relatively formal. It is noteworthy that to each regularized product we associate naturally two nonnegative integers determined directly from the definition. Then the fundamental estimates of Chapter I show that the order of growth of the logarithmic derivatives of such products in strips is determined by these two integers. In the application to the evaluation of certain integrals involving test functions, one can then see that the order of decay of these test functions, needed to insure that the integrals converge, is also determined by these two integers. Our systematic approach both improves known estimates for the Selberg zeta function (cf. Chapter I, §4), and provides estimates for functions in our class which had not been considered previously.

Theta inversions. We shall postpone to still another work the application of explicit formulas to the counting of those objects which play the role of prime powers. Here we shall emphasize an entirely different type of application, obtained by taking Gaussian type functions as the test functions instead of other test functions which lead to the counting. Applying the general explicit formula to such Gaussians gives rise to relations which are vast generalizations of the classical Jacobi inversion formula for the classical Jacobi theta function, where t on one side gets inverted to $1/t$ on the other side of the formula. The classical Jacobi inversion formula is the relation

$$\frac{1}{2\pi} \sum_{n=-\infty}^{\infty} e^{-n^2 t} = \frac{1}{\sqrt{4\pi t}} \sum_{n=-\infty}^{\infty} e^{-(2\pi n)^2/4t},$$

which holds for all $t > 0$. Here, $\log \mathbf{q} = 2\pi n$ where n is a positive integer. The zeta function $Z(s)$ giving rise to the above theta series is essentially the special Dirichlet polynomial

$$\sin(\pi i s) = -\frac{1}{2i} e^{\pi s}(1 - e^{-2\pi s}).$$

Thus, the most classical theta series appears in a new context, associated to a "zeta function" which looks quite different from those visualized classically.

The general context of Chapter IV and Chapter V allows a formulation of a theta inversion when the theta series is of type

$$\sum_k a_k e^{-\lambda_k t}$$

with various coefficients a_k. Theta inversion applies in certain cases when the sequence $\{\lambda_k\}$ is the sequence of eigenvalues of an operator. For example, as we will show in Chapter V, §4, such an inversion formula comes directly from considering the heat kernel on the compact quotient of an odd dimensional hyperbolic space which has metric with constant negative sectional curvature.

For certain manifolds, the theta inversion already gives rise to an extended class of zeta functions, which instead of an Euler sum may have a Bessel sum. For manifolds of even dimension, the class of functions having an Euler sum or Bessel sum is still not adequate, and it is necessary to define an even further extended class, which we shall describe briefly below. At this moment, it is not yet completely clear just how far an extension we shall need, but so far, whatever the extension of the fundamental class we have met, the techniques of [JoL 93a,b,c] and of Chapter I apply.

In [JoL 94], we show how the general explicit formula also applies to the scattering determinant of Eisenstein series. Here, the Euler sum exists, and scattering determinants are in the fundamental class.

An additive theory rather than multiplicative theory, and an extended class of functions. The conditions defining our fundamental class of functions are phrased in a manner still relatively close to the classical manner, involving the functions multiplicatively. However, it turns out that many essential properties of these functions involve only their logarithmic derivative, and thus give rise to an additive theory. For a number of applications, it is irrelevant that the residues are integers, and in some applications we are forced to deal with the more general notions of a regularized harmonic series (suitably normalized Mittag-Leffler expansions, with poles of order one) whose definition is recalled in Chapter I, §1. In general, the residues of such a series are not integers, so one cannot integrate back to realize this series as a logarithmic derivative of a meromorphic function. Even for the Artin L-functions, although they can be defined by an Euler product, it was natural for Artin to define them via their logarithmic derivative, and at the time, Artin could only prove that the residues were rational numbers. It took many years before the residues were finally proved to be integers. The systematic approach of [JoL 93a,b,c] in fact has been carried out so that it applies to this additive situation. The example of Chapter V, §4, shows why such an additive theory is essential.

Thus we are led to define not only the fundamental class of functions whose logarithmic derivative admits a Dirichlet series expression as mentioned above, but an extended class of functions where this condition is replaced by another one which will allow applications to more situations, starting with applications to various spectral theories as in [JoL 94]. Nevertheless, we still defined the fundamental class of functions having Euler sums, and we phrase some results multiplicatively, partly because at the present time, we feel that a complete change of notation with existing works would only make the present work less accessible, and partly because the class of functions admitting Euler sums is still a very important one including the classical functions of algebraic number theory and representation theory. However, we ask readers to keep in mind the additive rather than multiplicative formalism. Many sections, e.g. Chapter I and §1 and §2 of Chapter V, are written so that they apply directly to the additive situation.

Functions in the multiplicative fundamental class are obtained as Mellin transforms of theta functions having an inversion formula. Functions in the extended additive class are obtained as regularized harmonic series which are Gaussian transforms of such theta functions. For example, the (not regularized) harmonic series obtained from the heat kernel theta function in the special case of compact quotients of the three dimensional, complete, simply connected, hyperbolic manifold is essentially

$$\sum_k \frac{\phi_k(x)\phi_k(y)}{s(s-2)+\lambda_k}.$$

Observe how the presence of $s(s-2)$ in the series formally insures a trivial functional equation, that is invariance under $s \mapsto 2-s$.

Conversely, given a function in our extended additive class, one may go in reverse and see that the original theta inversion is only a special case of the general explicit formula valid for much more general test functions. The existence of an explicit formula with a more general test function will then allow us to obtain various counting results in subsequent publications.

Finally, let us note that many examples of explicit formulas using various test functions involving many examples of zeta functions have been treated in the literature, providing a vast number of papers on the subject. Most of the papers dealing with such explicit formulas are not directly relevant for what we do here, which is to lay out a general inductive "ladder principle" for explicit formulas in line with our treatment of Cramér's theorem. For instance, Deninger in [Den 93] emphasizes the compatibility of an explicit formula for the Riemann zeta function with a conjectural formalism of a Lefschetz trace formula. Such a formalism might occur in the presence of an operator whose eigenvalues are zeros of the zeta function. Our inductive hypotheses cover a wider class of functions than in [Den 93], and our treatment emphasizes another direction in the study of regularized products and series. Factors of regularized product type behave as if there were an operator, but no operator may be available.

We also mention Gallagher's attempt to unify a treatment of Selberg's trace formula with treatments of ordinary analytic number theory [Ga 84]. However, the conditions under which Gallagher proves his results are very restrictive compared to ours, and, in particular, are too restrictive to take into account the inductive ladder principle which we are following.

Acknowledgement: During the preparation of this work, the first author received support from NSF grant DMS-93-07023. Both authors benefited from visits to the Max-Planck-Institut in Bonn.

CHAPTER I

Asymptotic estimates of regularized harmonic series.

The proof of the general explicit formulas for functions whose fudge factors are of regularized product type will require a number of asymptotic estimates of general regularized harmonic series. The purpose of this chapter is to establish and tabulate these estimates in convenient form. The main definitions and results of this chapter are stated in §1 and §6.

These asymptotic formulas are needed just as one needs the asymptotic behavior of the gamma function and the zeta function in classical analytic number theory (see, for example, Chapter XVII of [La 70]). However, classical arguments which estimate this behavior cannot be applied in general, and must be replaced by more powerful tools, such as our extension of Cramér's theorem, proved in [JoL 93c], as well as our systematic analysis of the regularized harmonic series, given in [JoL 93a] and [JoL 93b].

Following the notation of [JoL 93a] and [JoL 93b], we let $R(z)$ be the regularized harmonic series associated to the theta function $\theta(t) = \sum a_k e^{-\lambda_k t}$; in other words

$$R(z) = \mathrm{CT}_{s=1} \mathbf{LM}\theta(s, z)$$

where $\mathbf{LM}$ is the Laplace-Mellin transform, and $\mathrm{CT}_{s=1}$ is the constant term of the power series in s at $s = 1$. As is shown in §4 of [JoL 93a], the function $R(z)$ has a meromorphic continuation to all $z \in \mathbf{C}$ whose singularities are simple poles located at $z = -\lambda_k$ with corresponding residue a_k. In what we call **the spectral case**, meaning $a_k \in \mathbf{Z}$ for all k, we have a regularized product $D(z)$ which is a meromorphic function defined for all $z \in \mathbf{C}$ and which satisfies the relation

$$R(z) = D'/D(z).$$

However, in this chapter, we will work in the more general situation by considering a regularized harmonic series which is not necessarily the logarithmic derivative of a regularized product. Our basic tool for estimating $R(z)$ is our general Gauss formula, which we shall recall at the end of §1.

From the general Gauss formula, we shall determine the asymptotic behavior of $R(z)$ as $z \to \infty$ in each of the following cases:

1) in a vertical strip obtained by restricting $\mathrm{Re}(z)$ to a compact interval;
2) in a sector $|\mathrm{Im}(z)| \ll \mathrm{Re}(z)$;
3) in a sequence of vertical line segments $z = -T_m + iy$ with $T_m \to \infty$ and y in a compact interval so that $R(z)$ grows as slowly as possible;
4) in a strip parallel and disjoint from a strip which contains all $-\lambda_k$, assuming that $\lambda_k \to \infty$ in a horizontal strip.

In all cases, the asymptotic behavior of $R(z+w)$ will be determined from the general Gauss formula through judicious choices of z and w.

In §6 we apply these results to functions which are obtained from regularized products by a suitable change of variables $z \mapsto \alpha z + \beta$ with $\alpha, \beta \in \mathbf{C}$. Such a change of variables is needed, for example, because zeta functions usually have their zeros in vertical strips and not in horizontal strips. Power products of regularized products subject to such changes of variables will be said to be of regularized product type, formally defined in §6.

We conclude this chapter by comparing our results to various examples that exist elsewhere in the literature.

Recall that a regularized harmonic series has a natural "reduced order M" which is closely related to the exponent of the leading term in the asymptotic expansion of the associated theta function near $t = 0$. We recall the precise definition at the end of §1. There is another characterization of the reduced order in the spectral case, since we can then write the regularized product as a Weierstrass product

$$D(z) = e^{P(z)} E(z)$$

where P has degree $m+1$ and E is a canonical Weierstrass product of order $\leq m$. The smallest integer m for which such an expression is possible is the reduced order of D. In the classical case of analytic number theory, the gamma and zeta functions of number fields have reduced order 0.

From §1 to §5, we let R be a regularized harmonic series whose definition will be recalled in §1, as well as other basic definitions used throughout the chapter.

§1. Regularized products and harmonic series.

Let us briefly recall necessary background material from the theory of regularized products and series, as established in [JoL 93a] and [JoL 93b], to which we refer for details and further results.

We let $L = \{\lambda_k\}$ and $A = \{a_k\}$ be sequences of complex numbers which may be subject to the following conditions.

DIR 1. For every positive real number c, there is only a finite number of k such that $\mathrm{Re}(\lambda_k) \leq c$.

We use the convention that $\lambda_0 = 0$ and $\lambda_k \neq 0$ for $k \geq 1$. Under condition **DIR 1** we delete from the complex plane $\mathbf{C}$ the horizontal half lines going from $-\infty$ to $-\lambda_k$ for each k, together, when necessary, the horizontal half line going from $-\infty$ to 0. We define the open set:

$\mathbf{U}_L$ = the complement of the above half lines in $\mathbf{C}$.

If all λ_k are real and positive, then we note that $\mathbf{U}_L$ is simply $\mathbf{C}$ minus the negative real axis $\mathbf{R}_{\leq 0}$.

DIR 2.
   (a) The Dirichlet series

$$\sum_k \frac{a_k}{\lambda_k^\sigma}$$

   converges absolutely for some real σ, say σ_0.

   (b) The Dirichlet series

$$\sum_k \frac{1}{\lambda_k^\sigma}$$

   converges absolutely for some real σ, say σ_1.

DIR 3. There is a fixed $\epsilon > 0$ such that for all k sufficiently large, we have

$$-\frac{\pi}{2} + \epsilon \leq \arg(\lambda_k) \leq \frac{\pi}{2} - \epsilon.$$

We will consider a **theta series** or **theta function**, which is defined by

$$\theta_{A,L}(t) = \theta(t) = a_0 + \sum_{k=1}^{\infty} a_k e^{-\lambda_k t},$$

and, for each integer $N \geq 1$, we define the **asymptotic exponential polynomials** by

$$Q_N(t) = a_0 + \sum_{k=1}^{N-1} a_k e^{-\lambda_k t}.$$

We are also given a sequence of complex numbers $\{p\} = \{p_j\}$ with

$$\mathrm{Re}(p_0) \leq \mathrm{Re}(p_1) \leq \cdots \leq \mathrm{Re}(p_j) \leq \ldots$$

increasing to infinity, and, to every p in this sequence, we associate a polynomial B_p of degree n_p and set

$$b_p(t) = B_p(\log t).$$

We then define the **asymptotic polynomials at** 0 by

$$P_q(t) = \sum_{\mathrm{Re}(p)<\mathrm{Re}(q)} b_p(t) t^p.$$

We define

$$m(q) = \max \deg B_p \quad \text{for} \quad \mathrm{Re}(p) = \mathrm{Re}(q)$$

$$n(q) = \max \deg B_p \quad \text{for} \quad \mathrm{Re}(p) < \mathrm{Re}(q),$$

$$n(q') = \max \deg B_p \quad \text{for} \quad \mathrm{Re}(p) \leq \mathrm{Re}(q).$$

We shall use the term **special case** to describe the instance when $n(q) = 0$ for all q. The **principal part** of $\theta(t)$ is defined to be

$$P_0\theta(t) = \sum_{\mathrm{Re}(p)<0} b_p(t) t^p.$$

Let $\mathbf{C}\langle T \rangle$ be the algebra of polynomials in T^p with arbitrary complex powers $p \in \mathbf{C}$. Then, with this notation, $P_q(t) \in \mathbf{C}[\log t]\langle t \rangle$.

The function θ on $(0, \infty) = \mathbf{R}_{>0}$ is subject to **asymptotic conditions:**

AS 1. Given a positive number C and $t_0 > 0$, there exists N and $K > 0$ such that

$$|\theta(t) - Q_N(t)| \leq Ke^{-Ct} \text{ for } t \geq t_0.$$

AS 2. For every q, we have

$$\theta(t) - P_q(t) = O(t^{\mathrm{Re}(q)} |\log t|^{m(q)}) \text{ for } t \to 0,$$

which shall written as

$$\theta(t) \sim \sum_p b_p(t) t^p.$$

AS 3. Given $\delta > 0$, there exists an $\alpha > 0$ and a constant $C > 0$ such that for all N and $0 < t \leq \delta$ we have

$$|\theta(t) - Q_N(t)| \leq C/t^\alpha.$$

We shall assume throughout that the theta series converges absolutely for $t > 0$. From **DIR 1** it follows that the convergence of the theta series is uniform for $t \geq \delta > 0$ for every δ.

The **Laplace-Mellin transform** of a measurable function f on $(0, \infty)$ is defined by

$$\mathbf{LM}f(s, z) = \int_0^\infty f(t) e^{-zt} t^s \frac{dt}{t}.$$

Theorem 1.1. *Let θ satisfy* **AS 1**, **AS 2** *and* **AS 3**. *Then* **LM**θ *has a meromorphic continuation for $s \in \mathbf{C}$ and $z \in \mathbf{U}_L$. For each z, the function $s \mapsto \mathbf{LM}\theta(s, z)$ has poles only at the points $-(p+n)$ with $b_p \neq 0$ in the asymptotic expansion of θ at 0. A pole at $-(p+n)$ has order at most $n(p') + 1$. In the special* case when the asymptotic expansion at 0 has no log terms, the poles are simple.

We shall use a systematic notation for the coefficients of the Laurent expansion of $\mathbf{LM}\theta(s,z)$ near $s = s_0$. Namely we let $R_j(s_0; z)$ be the coefficient of $(s - s_0)^j$, so that

$$\mathbf{LM}\theta(s,z) = \sum R_j(s_0; z)(s - s_0)^j.$$

The constant term $R_0(s_0; z)$ is so important that we give it a special notation, namely,

$$\mathbf{CT}_{s=s_0} \mathbf{LM}\theta(s,z) = R_0(s_0; z).$$

In particular, we define the **regularized harmonic series** $R(z)$ to be the meromorphic function defined by

$$R(z) = \mathbf{CT}_{s=1} \mathbf{LM}\theta(s,z) = R_0(1; z).$$

Theorem 1.2. Let θ satisfy **AS 1**, **AS 2** and **AS 3**.
 a) For every $z \in \mathbf{U}_L$ and s near 0, the function $\mathbf{LM}\theta(s,z)$ has a pole at $s = 0$ of order at most $n(0') + 1$, and the function $\mathbf{LM}\theta(s,z)$ has the Laurent expansion

$$\mathbf{LM}\theta(s,z) = \frac{R_{-n(0')-1}(0; z)}{s^{n(0')+1}} + \cdots + R_0(0; z) + R_1(0; z)s + \ldots$$

 where, for each $j < 0$, $R_j(0; z) \in \mathbf{C}[z]$ is a polynomial of degree $\leq -\mathrm{Re}(p_0)$.
 b) One has the differential equation

$$\partial_z \mathbf{LM}\theta(s,z) = -\mathbf{LM}\theta(s+1,z),$$

and hence the relation

$$-\partial_z R_0(0; z) = R(z).$$

The regularized harmonic series is a particular meromorphic function that has simple poles at $z = -\lambda_k$ with residue a_k.

Next we recall conditions when $R(z)$ is the logarithmic derivative of a meromorphic function. We define the **spectral case** to be when all $a_k \in \mathbf{Z}$.

Theorem 1.3. *In the spectral case, there exists a unique mero-morphic function $D(z)$, called the **regularized product**, such that*

$$- \log D(z) = \mathrm{CT}_{s=0} \mathbf{LM}\theta(s, z) = R_0(0; z).$$

We have the relation

$$D'/D(z) = R(z).$$

This regularized product is meromorphic of finite order having zeros at the points $z = -\lambda_k$ with multiplicity a_k.

To make the situation more explicit, and to compare with results that exist elsewhere in the literature, let us record the following formula. In the spectral case, define

$$\zeta_\theta(s, z) = \frac{1}{\Gamma(s)} \mathbf{LM}\theta(s, z)$$

If we assume that

$$\theta(t) = \sum_{k=1}^{\infty} a_k e^{-\lambda_k t}$$

satisfies the asymptotic conditions **AS 1**, **AS 2** and **AS 3**, then we have the equality

$$\zeta_\theta(s, z) = \sum_{k=1}^{\infty} \frac{a_k}{(z + \lambda_k)^s}$$

for $\mathrm{Re}(z)$ and $\mathrm{Re}(s)$ sufficiently large. By Theorem 1.1 and Theorem 1.2, $\zeta_\theta(s, z)$ is holomorphic at $s = 0$ for $z \in \mathbf{U}_L$ and

$$\zeta_\theta'(0, z) = \mathrm{CT}_{s=0} \mathbf{LM}\theta(s; z) + \gamma R_{-1}(0; z) = R_0(0; z) + \gamma R_{-1}(0; z).$$

Thus, in the spectral and special case, $-R_0(0; z)$ amounts to a normalization of the analytic torsion $-\zeta'(0, z)$. We have the Lerch formula

$$D'/D(z) = -\zeta'(0, z) + \quad \text{a constant.}$$

However, even in the most general case, for many applications one can work just as well with the regularized harmonic series $R(z)$ or

$R_0(0; z)$ even through $\zeta(s, z)$ is not holomorphic at $s = 0$, and D does not exist. For further comments, see §4 of Chapter V.

For any regularized harmonic series R, or regularized product D in the spectral case, we define the **reduced order** to be the pair of integers (M, m) where, in the notation of the asymptotic condition **AS 2**:

M is the largest integer $< -\text{Re}(p_0)$;

$m = m(p_0) + 1$ if there is a complex index p with

$$\text{Re}(p_0) = \text{Re}(p) \in \mathbf{Z}_{<0} \quad \text{and} \quad b_p \neq 0,$$

otherwise simply set $m = m(p_0)$.

Finally we recall the Gauss formula of [JoL 93a], §4. For any complex index q with $\text{Re}(q) > 0$, and variables z and w with $\text{Re}(w) > 0$ and

$$\text{Re}(w) > -\text{Re}(z) - \text{Re}(\lambda_k) \quad \text{for all } k$$

we then have the equality

$$R(z + w) = I_w(z; q) + S_w(z; q)$$

where

$$I_w(z; q) = \int_0^\infty [\theta_z(t) - P_q \theta_z(t)] \, e^{-wt} dt,$$

with

$$\theta_z(t) = e^{-zt} \theta(t)$$

and

$$S_w(z; q) = \sum_{\text{Re}(p)+k<\text{Re}(q)} \frac{(-z)^k}{k!} \text{CT}_{s=0} B_p(\partial_s) \left[\frac{\Gamma(s + p + k + 1)}{w^{s+p+k+1}} \right].$$

§2. Asymptotics in vertical strips.

In this section we will determine the asymptotic behavior of a regularized harmonic series in a vertical strip. We shall see that the asymptotics in a vertical strip are given by the term $S_w(z; q)$.

We let

$$\Lambda_n = \max_{k \geq n}\{-\mathrm{Re}(\lambda_k)\}.$$

Note that Λ_n is bounded above for all n and $\Lambda_n \to -\infty$ as $n \to \infty$.

Theorem 2.1. *Let $x_1, x_2 \in \mathbf{R}$ with $x_1 < x_2$. Select n sufficiently large so that $\Lambda_n < x_1 + 1$. Then for all q with $\mathrm{Re}(q) > 0$ and uniformly for $x \in [x_1, x_2]$, we have the asymptotic relation*

$$R(x+1+iy) = S_{1+iy}(x; q) + \sum_{k=0}^{n-1} \frac{a_k}{x+1+iy+\lambda_k} + o\left(|y|^{-[\mathrm{Re}(q)]}\right)$$

for $|y| \to \infty$.

Proof. With n chosen as above, let $L_n = \{\lambda_n, \dots\}$. We decompose $R(z)$ as

$$R(z) = \sum_{k=0}^{n-1} \frac{a_k}{z+\lambda_k} + R_n(z),$$

where $R_n(z)$ is the regularized harmonic series formed from the sequence Λ_n with coefficients $\{a_k\}$, $k \geq n$. The desired estimate is obvious for the finite sum, so we may assume that we are working with the sequence L_n, and we will suppress this subscript. We apply the general Gauss formula by setting $z = x$ and $w = 1 + iy$ with $y \in \mathbf{R}$. It suffices to prove the estimate

$$I_{1+iy}(x; q) = o(|y|^{-\mathrm{Re}(q)}) \qquad \text{for } |y| \to \infty.$$

The integer n has been chosen sufficiently large so that, as a function of t,

$$\theta_x(t)e^{-t} = O\left(e^{-(x_1+1-\Lambda_n)t}\right) \qquad \text{for } x \in [x_1, x_2] \text{ and } t > 1.$$

In particular, this implies, by **AS 1** and **DIR 2**, that we have

$$\theta_x(t)e^{-t} \in L^1[1, \infty) \cap C^\infty[1, \infty).$$

Since $P_q \theta_x(t)$ has polynomial growth in t,

$$P_q \theta_x(t) e^{-t} \in L^1[1, \infty) \cap C^\infty[1, \infty).$$

Directly from **AS 2**, we have

$$[\theta_x(t) - P_q \theta_x(t)] e^{-t} \in L^1[0, 1] \cap C^{[\mathrm{Re}(q)]}[0, 1],$$

and

$$[\theta_x(t) - P_q \theta_x(t)] e^{-t} = O(t^{\mathrm{Re}(q)} |\log t|^{m(q)}) \quad \text{for } t \to 0.$$

We now apply the Riemann-Lebesgue lemma to obtain the bound

$$I_{1+iy}(x; q) = \int\limits_0^\infty [\theta_x(t) - P_q \theta_x(t)] e^{-t} e^{-iyt} dt = o(|y|^{-[\mathrm{Re}(q)]})$$

which holds uniformly for all $x \in [x_1, x_2]$ as $|y| \to \infty$. With this, the proof of the theorem is complete. $\quad\square$

For many purposes, it suffices simply to know the lead asymptotic of $R(z)$, which is obtained from the bound

$$\sum_{k=0}^{n-1} \frac{a_k}{x + 1 + iy + \lambda_k} = o(1) \quad \text{for } |y| \to \infty,$$

and the following result.

Corollary 2.2. *Let* $m = m(p_0) + 1$ *if there is an index p with* $\mathrm{Re}(p_0) = \mathrm{Re}(p) \in \mathbf{Z}_{<0}$ *and* $b_p \neq 0$, *otherwise set* $m = m(p_0)$. *Then for any q with* $\mathrm{Re}(q) > 0$, *we have*

$$S_{1+iy}(x; q) = O\left(|y|^{-\mathrm{Re}(p_0)-1}(\log |y|)^m\right) \quad \text{for } |y| \to \infty.$$

Since $-\mathrm{Re}(p_0) - 1 \leq M < -\mathrm{Re}(p_0)$, *this estimate can be written as*

$$S_{1+iy}(x; q) = O\left(|y|^M (\log |y|)^m\right) \quad \text{for } |y| \to \infty.$$

Proof. Immediate from the definition of $S_w(z; q)$ as applied to $S_{1+iy}(x; q)$. $\quad\square$

§3. Asymptotics in sectors.

In this section we will determine the asymptotic behavior of the regularized harmonic series $R(w)$ as $w \to \infty$ in a sector of the form

$$\text{Sec}_\epsilon = \{w \in \mathbf{C} \mid -\frac{\pi}{2} + \epsilon < \arg(w) < \frac{\pi}{2} - \epsilon\}$$

for some $\epsilon > 0$. As in the previous section, the asymptotics are determined by the term $S_w(0; q)$ in the general Gauss formula.

Theorem 3.1. *Let* $x = \text{Re}(w)$. *For all* q *with* $\text{Re}(q) > 0$, *we have*

$$R(w) = S_w(0; q) + O(x^{-\text{Re}(q)-1}(\log x)^{m(q)})$$

as $w \to \infty$ *in* Sec_ϵ *where, as above,*

$$S_w(0; q) = \sum_{\text{Re}(p) < \text{Re}(q)} \text{CT}_{s=0} B_p(\partial_s) \left[\frac{\Gamma(s + p + 1)}{w^{s+p+1}} \right].$$

Proof. We apply the general Gauss formula with $z = 0$. The proof follows from estimating the integrals

$$I_w(0; q) = \int_0^1 [\theta(t) - P_q\theta(t)] e^{-wt} dt + \int_1^\infty [\theta(t) - P_q\theta(t)] e^{-wt} dt.$$

By **AS 2**, we have

$$\left| \int_0^1 [\theta(t) - P_q\theta(t)] e^{-wt} dt \right| \ll \int_0^1 t^{\text{Re}(q)} |\log t|^{m(q)} e^{-xt} dt$$

$$\leq \left[(\partial_s)^{m(q)} \frac{\Gamma(s + 1)}{x^{s+1}} \right]_{s=\text{Re}(q)} + \int_1^\infty t^{\text{Re}(q)} (\log t)^{m(q)} e^{-xt} dt$$

$$= O\left(x^{-\text{Re}(q)-1}(\log x)^{m(q)} \right)$$

as $w \to \infty$ in Sec_ϵ. For $t \geq 1$, $\theta(t) - P_q\theta(t) = O(e^{ct})$ for some $c > 0$ and so, for $w \in \mathrm{Sec}_\epsilon$ with x sufficiently large, we have

$$\left| \int_1^\infty [\theta(t) - P_q\theta(t)] \, e^{-wt} \, dt \right| \leq K \int_1^\infty e^{-(x-c)t} dt = O\left(e^{-x}/x\right),$$

thus yielding the stated estimate. $\square$

As in the case of asymptotics in vertical strips, one often needs to know only the lead asymptotics of $R(z)$. Using Corollary 2.2, we can state this result.

Corollary 3.2. *Let $m = m(p_0) + 1$ if there is an index p with $\mathrm{Re}(p_0) = \mathrm{Re}(p) \in \mathbf{Z}_{<0}$ and $b_p \neq 0$, otherwise set $m = m(p_0)$. Then, with notation as above,*

$$R(w) = O\left(x^{-\mathrm{Re}(p_0)-1}(\log x)^m\right) \quad \text{for } w \to \infty \text{ in } \mathrm{Sec}_\epsilon.$$

In particular, since $-\mathrm{Re}(p_0) - 1 \leq M < -\mathrm{Re}(p_0)$, this estimate can be written as

$$R(w) = O\left(x^M (\log x)^m\right) \quad \text{for } w \to \infty \text{ in } \mathrm{Sec}_\epsilon.$$

It is interesting to note that the asymptotic relation given in Corollary 3.2 is identical to the result given in Corollary 2.2 even though the directions in which $w \to \infty$ are quite different.

We can integrate the result in Theorem 3.1 in order to obtain an alternative proof of the general Stirling's formula for a regularized product, first established in §5 of [JoL 93a]. For completeness, we shall simply state this result.

Theorem 3.3. *Let D be a regularized product, so that we have $R = D'/D$. Let*

$$\mathbf{B}_q(w) = \sum_{\mathrm{Re}(p) < \mathrm{Re}(q)} \mathrm{CT}_{s=0} B_p(\partial_s) \left[\frac{\Gamma(s+p)}{w^{s+p}} \right] \in \mathbf{C}\langle w \rangle [\log w],$$

and set $x = \mathrm{Re}(w)$. Then for all q with $\mathrm{Re}(q) > 0$, we have

$$\log D(w) = \mathbf{B}_q(w) + O(x^{-\mathrm{Re}(q)}(\log x)^{m(q)})$$

as $w \to \infty$ in Sec_ϵ.

§4. Asymptotics in a sequence to the left.

In this section we will determine the asymptotic behavior of the regularized harmonic series $R(z)$ for a particular sequence of vertical line segments going to the left so that $R(z)$ grows as slowly as possible. Roughly, these line segments must pass between certain consecutive pairs of poles of $R(z)$ that are "sufficiently far apart".

We shall state and prove results for general regularized harmonic series and for regularized harmonic series which are the logarithmic derivatives of regularized products. The main result of this section is the following theorem.

Theorem 4.1. *There is a sequence of real numbers $T_n \to \infty$ such that for all y in any compact interval of $\mathbf{R}$ the following asymptotic relations hold.*

a) *In the notation of* **DIR 2**, *we have*

$$R(-T_n + iy) = o(T_n^{\sigma_1 + \sigma_0 - 1} \log T_n) \quad \text{as } T_n \to \infty.$$

b) *Let* $m = m(p_0) + 1$ *if there is a complex index p with* $\mathrm{Re}(p_0) = \mathrm{Re}(p) \in \mathbf{Z}_{<0}$ *and $b_p \neq 0$, otherwise simply set* $m = m(p_0)$. *Then for any q with $\mathrm{Re}(q) > 0$, we have*

$$R(-T_n + iy) = O(T_n^{-\mathrm{Re}(p_0)-1}(\log T_n)^m) \quad \text{as } T_n \to \infty.$$

In particular, since $-\mathrm{Re}(p_0) - 1 \le M < -\mathrm{Re}(p_0)$, *we have*

$$R(-T_n + iy) = O(T_n^M (\log T_n)^m) \quad \text{as } T_n \to \infty.$$

The proof of Theorem 4.1 comes from considering the general Gauss formula for $R(u + w)$ with

$$w = -T - 1 + iy \quad \text{and} \quad u \in [0, 1],$$

for various choices of T and u in $\mathbf{R}$ which are to be made later. For T sufficiently large, let us write

$$(1) \quad R(u + w) = \sum_{|\lambda_k| \le 2T} \frac{a_k}{u + w + \lambda_k} + S_w(u; 0) + I_w^{(2T)}(u; 0),$$

where

$$\theta_u^{(2T)}(t) = \sum_{|\lambda_k|>2T} a_k e^{-\lambda_k t} \cdot e^{-ut}$$

$$= \theta_u(t) - \sum_{|\lambda_k|\leq 2T} a_k e^{-\lambda_k t} \cdot e^{-ut}$$

and

$$I_w^{(2T)}(u;0) = \int_0^\infty \left[\theta_u^{(2T)}(t) - P_0 \theta_u^{(2T)}(t) \right] e^{-wt} dt.$$

In (1) we have used the relation

$$P_0 \theta_u^{(2T)}(t) = P_0 \theta_u(t),$$

which, in particular, gives the identity

$$S_w^{(2T)}(u;0) = S_w(u;0).$$

Theorem 4.1 will follow by estimating each term in (1) separately. In a manner similar to that of Corollary 2.2, we can estimate the term $S_w(u;0)$. Specifically, with the above choice of variables u and $w = -T - 1 + iy$, we immediately have

$$(2) \qquad S_w(u;0) = O\left(T^{-\mathrm{Re}(p_0)-1}(\log T)^m\right) \qquad \text{for } |T| \to \infty.$$

This bound fits the result stated in Theorem 4.1(a) after noting the relations

$$\sigma_0 > 0 \quad \text{and} \quad \sigma_1 > -\mathrm{Re}(p_0).$$

Throughout this section, which is devoted to the proof of Theorem 4.1, we will consider both general regularized harmonic series and those regularized harmonic series which are logarithmic derivatives of regularized products. In this latter case, we will, for this section only, assume that all coefficients a_k are equal to ± 1 and count the zeros λ_k with multiplicities.

In order to estimate the remaining two terms in (1), we need the following bounds.

Lemma 4.2. *With the notation as above, the following asymptotic relations hold:*

a) *In the notation of* **DIR 2**, *we have*

$$k = o(|\lambda_k|^{\sigma_1}) \quad \text{and} \quad |a_k| = o(|\lambda_k|^{\sigma_0}) = o\left(k^{\sigma_0/\sigma_1}\right).$$

b) *In the spectral case, there is a positive constant C' such that*

$$k \sim C' |\lambda_k|^{-\text{Re}(p_0)} (\log |\lambda_k|)^{m(p_0)},$$

or, equivalently, there is a positive constant C such that

$$|\lambda_k| \sim C \left(k/(\log k)^{m(p_0)}\right)^{-1/\text{Re}(p_0)}.$$

Proof. Define the **absolute zeta function** associated to the sequence L by

$$\zeta_{\text{abs}}(s) = \sum_{k=1}^{\infty} |\lambda_k|^{-s}.$$

By the inverse Mellin transform (see §7 of [JoL 93a]), we have, for all $\sigma > \sigma_1$

$$t^{\sigma_1} \cdot \sum_{k=1}^{\infty} e^{-|\lambda_k|t} = \frac{1}{2\pi i} \int_{\sigma-i\infty}^{\sigma+i\infty} \zeta_{\text{abs}}(s)\Gamma(s)t^{-(s-\sigma_1)}ds.$$

Since $\text{Re}(-s + \sigma_1) < 0$, we can apply the dominated convergence theorem to conclude

$$(3) \qquad \lim_{t \to 0}\left[t^{\sigma_1} \cdot \sum_{k=1}^{\infty} e^{-|\lambda_k|t}\right] = 0.$$

Similarly, one shows that

$$(4) \qquad \lim_{t \to 0}\left[t^{\sigma_0} \cdot \sum_{k=1}^{\infty} |a_k| e^{-|\lambda_k|t}\right] = 0.$$

In the spectral case, we have, by **AS 2**, the existence of a positive constant c such that

$$(5) \qquad \lim_{t \to 0}\left[t^{-\text{Re}(p_0)}(\log t)^{-m(p_0)} \cdot \sum_{k=1}^{\infty} e^{-|\lambda_k|t}\right] = c.$$

At this point, we apply the following general result.

The Karamata Theorem. *Let $\alpha \in \mathbf{R}_{>0}$ and $\beta \in \mathbf{Z}_{\geq 0}$. Let μ be a positive measure on $\mathbf{R}^+$ such that the integral*

$$\int_0^\infty e^{-t\lambda} \mu(\lambda)$$

converges for $t > 0$, and such that

$$\lim_{t \to 0} \left[t^\alpha (-\log t)^\beta \int_0^\infty e^{-t\lambda} \mu(\lambda) \right] = C$$

for some positive constant C. If $f(x)$ is a continuous function on the interval $[0,1]$, then

$$\lim_{t \to 0} \left[t^\alpha (-\log t)^\beta \int_0^\infty f(e^{-t\lambda}) e^{-t\lambda} \mu(\lambda) \right]$$

$$= \frac{C}{\Gamma(\alpha)} \int_0^\infty f(e^{-t}) t^\alpha e^{-t} \frac{dt}{t}.$$

The proof of the Karamata theorem in the above form with the factor $(-\log t)^\beta$ follows the proof given on page 94 of [BGV 92].

We apply the Karamata theorem as on page 95 of [BGV 92], namely by considering a decreasing sequence of continuous test functions converging to the function x^{-1} on the interval $[1/e, 1]$ and zero on the interval $[0, 1/e)$. In the setup of (a), we use the spectral measure that places unit mass at the points $|\lambda_k| \in \mathbf{R}^+$, taking into account multiplicities. From (3), we obtain the bound

(6) $$N(\lambda) = \#\{k : |\lambda_k| \leq \lambda\} = o(\lambda^{\sigma_1}).$$

Since

$$N(\lambda_k) \geq k,$$

and we conclude

(7) $$k = o(|\lambda_k|^{\sigma_1}),$$

as was claimed in the first assertion of part (a).

To continue, let us again apply the Karamata theorem, this time with the measure that places a mass of $|a_k|$ at the points $|\lambda_k| \in \mathbf{R}^+$. The Karamata theorem then yields the estimate

$$\sum_{|\lambda_n| \leq \lambda} |a_n| = o(\lambda^{\sigma_0}),$$

which, in particular, implies the bound

$$|a_k| = o(|\lambda_k|^{\sigma_0}).$$

Combining this estimate with (7), we have

$$|a_k| = o(k^{\sigma_0/\sigma_1}),$$

which completes the proof of (a).

In the spectral case, we apply the Karamata theorem to (5) with the family of test functions described above and

$$\alpha = -\mathrm{Re}(p_0) \quad \text{and} \quad \beta = -m(p_0),$$

from which we obtain the estimate

(8) $$N(\lambda) = \#\{k : |\lambda_k| \leq \lambda\} \sim C'\lambda^\alpha (\log \lambda)^{-\beta}$$

for some positive constant C'. This estimate gives the relation

$$k = N(\lambda_k) \sim C'|\lambda_k|^\alpha (\log |\lambda_k|)^{-\beta},$$

as asserted in the statement of the lemma. Then

$$k^{1/\alpha} \sim C'^{1/\alpha}|\lambda_k|(\log |\lambda_k|)^{-\beta/\alpha}$$

and

$$\log k \sim \alpha \cdot \log |\lambda_k|.$$

Hence, for some positive constant C, we obtain the asymptotic relation

$$|\lambda_k| \sim C'^{-1/\alpha} k^{1/\alpha} (\log |\lambda_k|)^{\beta/\alpha} \sim C k^{1/\alpha} (\log k)^{\beta/\alpha},$$

which completes the proof of the lemma. $\square$

Lemma 4.2 allows us to estimate the term $I_w^{(2T)}(u; 0)$ from (1) in the following proposition.

Proposition 4.3. *Let $w = -T - 1 + iy$ and $u \in [0, 1]$. For all y in any compact interval of $\mathbf{R}$ the following asymptotic relations hold.*

a) *In the notation of* **DIR 2**, *we have*

$$I_w^{(2T)}(u; 0) = o\left(T^{\sigma_1 + \sigma_0 - 1} \log T\right) \quad \text{for } T \to \infty,$$

b) *Let $m = m(p_0) + 1$ if there is a p with $\mathrm{Re}(p_0) = \mathrm{Re}(p) \in \mathbf{Z}_{<0}$ and $b_p \neq 0$, otherwise set $m = m(p_0)$. Then, in the spectral case, we have*

$$I_w^{(2T)}(u; 0) = O\left(T^{-\mathrm{Re}(p_0) - 1}(\log T)^m\right) \quad \text{for } T \to \infty.$$

Proof. For n sufficiently large, one can write

$$\partial_u^n I_w^{(2T)}(u; 0) = (-1)^n \Gamma(n + 1) \sum_{|\lambda_k| > 2T} \frac{a_k}{(u + w + \lambda_k)^{n+1}}.$$

Since y lies in a compact interval, we have the estimate

$$|u + w| \sim T,$$

so we have the bound

$$\partial_u^n I_w^{(2T)}(u; 0) = O\left(\sum_{|\lambda_k| > 2T} \frac{|a_k|}{(|\lambda_k| - T)^{n+1}}\right).$$

Using Lemma 4.2(a) and an integral comparison, there exist positive constants c_1 and c_2 such that

$$\partial_u^n I_w^{(2T)}(u; 0) = o\left(\int_{c_1 T^{\sigma_1}}^{\infty} \frac{x^{\sigma_0/\sigma_1}}{(c_2 x^{1/\sigma_1} - T)^{n+1}} dx\right)$$

$$= o\left(\int_{c_1 T^{\sigma_1}}^{\infty} x^{\sigma_0/\sigma_1 - (n+1)/\sigma_1} dx\right)$$

$$= o\left(T^{\sigma_0 + \sigma_1 - (n+1)}\right).$$

To finish one integrates n-times. An extra power of $\log T$ occurs precisely in the case when $\sigma_0 + \sigma_1 \in \mathbf{Z}_{<0}$.

The proof of (b) is similar to that of (a). As above, let

$$\alpha = -\mathrm{Re}(p_0) \quad \text{and} \quad \beta = -m(p_0).$$

Lemma 4.2(b) and an integral comparison yield the estimate

$$\partial_u^n I_w^{(2T)}(u; 0) = O\left(\int_{c_1 T^\alpha (\log T)^{-\beta}}^\infty \frac{dx}{\left(c_2 \left(x(\log x)^\beta \right)^{1/\alpha} - T \right)^{n+1}} \right)$$

$$= O\left(\int_{c_1 T^\alpha (\log T)^{-\beta}}^\infty x^{-(n+1)/\alpha} (\log x)^{-\beta(n+1)/\alpha} dx \right).$$

If we integrate by parts, we obtain

$$\partial_u^n I_w^{(2T)}(u; 0) = O\left(\left. x^{-(n+1)/\alpha+1} (\log x)^{-\beta(n+1)/\alpha} dx \right|_{c_1 T^\alpha (\log T)^{-\beta}}^\infty \right)$$

$$= O\left(\left(T^\alpha (\log T)^{-\beta} \right)^{-(n+1)/\alpha+1} (\log T)^{-\beta(n+1)/\alpha} \right)$$

$$= O\left(T^{\alpha-(n+1)} (\log T)^{-\beta} \right)$$

$$= O\left(T^{-\mathrm{Re}(p_0)-(n+1)} (\log T)^{m(p_0)} \right).$$

To finish one integrates n-times. An extra power of $\log T$ occurs precisely in the case when $\mathrm{Re}(p_0) \in \mathbf{Z}_{<0}$. $\square$

The results stated in (2) and Proposition 4.3 are for $T \to \infty$. To estimate the remaining term in (1), namely the finite sum

$$\sum_{|\lambda_k| \leq 2T} \frac{a_k}{u + w + \lambda_k},$$

it is necessary to choose carefully a sequence T_n. Roughly speaking, the sequence must be as far from any $-\lambda_k$ as possible. The following lemma makes this statement precise.

Lemma 4.4. *Assume the notation as above.*

a) *Let* $r_n = |\lambda_n|^{\sigma_1}$. *There is a positive constant* c *and a sequence* $\{n_k\}$ *of positive integers with* $n_k \to \infty$ *such that*

$$r_{n_k+1} - r_{n_k} \geq c \quad \text{for all } n_k.$$

Equivalently, there is a positive constant c *and a sequence* $\{n_k\}$ *of positive integers with* $n_k \to \infty$ *such that*

$$|\lambda_{n_k+1}| - |\lambda_{n_k}| \geq c \cdot n_k^{1/\sigma_1 - 1} \quad \text{for all } n_k.$$

b) *Let* $r_n = |\lambda_n|^{\alpha}(\log|\lambda_n|)^{-\beta}$. *There is a positive constant* c *and a sequence* $\{n_k\}$ *of positive integers with* $n_k \to \infty$ *such that*

$$r_{n_k+1} - r_{n_k} \geq c \quad \text{for all } n_k.$$

Equivalently, there is a positive constant c *and a sequence* $\{n_k\}$ *of positive integers with* $n_k \to \infty$ *such that*

$$|\lambda_{n_k+1}| - |\lambda_{n_k}| \geq c \cdot n_k^{1/\alpha - 1}(\log n_k)^{-\beta/\alpha} \quad \text{for all } n_k.$$

Proof. For the first statement in (a), one has, by (6), the bound $k = o(r_k)$, which would be contradicted if no such constant c or subsequence $\{n_k\}$ would exist. The second assertion in (a) and both assertions in (b) are established similarly. $\square$

By assumption **DIR 3**, there is a constant C such that for all k, we have the inequalities

$$\text{Re}(\lambda_k) \leq |\lambda_k| \leq C\text{Re}(\lambda_k).$$

One then has results analogous to Lemma 4.2 and Lemma 4.4 for the sequence $\{\text{Re}(\lambda_k)\}$. Specifically, we shall need the following version of Lemma 4.4.

Lemma 4.5. *Assume the notation as above.*

a) *Let* $r_n = \text{Re}(\lambda_n)^{\sigma_1}$. *There is a positive constant* c *and a sequence* $\{n_k\}$ *of positive integers with* $n_k \to \infty$ *such that for all* n_k,

$$r_{n_k+1} - r_{n_k} \geq c.$$

Equivalently, there is a positive constant c and a sequence $\{n_k\}$ of positive integers with $n_k \to \infty$ such that for all n_k,

$$\operatorname{Re}(\lambda_{n_k+1}) - \operatorname{Re}(\lambda_{n_k}) \geq c \cdot n_k^{1/\sigma_1 - 1}.$$

b) Let $r_n = \operatorname{Re}(\lambda_n)^\alpha (\log \operatorname{Re}(\lambda_n))^{-\beta}$. There is a positive constant c and a sequence $\{n_k\}$ of positive integers with $n_k \to \infty$ such that for all n_k,

$$r_{n_k+1} - r_{n_k} \geq c.$$

Equivalently, there is a positive constant c and a sequence $\{n_k\}$ of positive integers with $n_k \to \infty$ such that for all n_k,

$$\operatorname{Re}(\lambda_{n_k+1}) - \operatorname{Re}(\lambda_{n_k}) \geq c \cdot n_k^{1/\alpha - 1} (\log n_k)^{-\beta/\alpha}.$$

The proof of Lemma 4.5 is identical to that of Lemma 4.4, hence will be omitted.

The following proposition estimates the finite sum in (1) for a particular sequence T_n of real numbers with $T_n \to \infty$ so that the finite sum grows as slowly as possible. With the proposition, the proof of Theorem 4.1 is complete.

Proposition 4.6. There is a sequence of real numbers T_n with $T_n \to \infty$ such that for all y in any compact interval of $\mathbf{R}$, the following asymptotic relations hold.

a) In the notation of **DIR 2**, we have

$$\sum_{|\lambda_k| \leq 2T_n} \frac{a_k}{-T_n + iy + \lambda_k} = o\left(T_n^{\sigma_1 + \sigma_0 - 1} \log T_n\right),$$

b) Let $m = m(p_0) + 1$. Then, in the spectral case, we have

$$\sum_{|\lambda_k| \leq 2T_n} \frac{a_k}{-T_n + iy + \lambda_k} = O\left(T_n^{-\operatorname{Re}(p_0) - 1} (\log T_n)^m\right).$$

Proof. For any z with $\text{Re}(z)$ sufficiently large and $|z| \neq |\lambda_k|$, we can estimate the finite sum in (a) by

$$\left| \sum_{|\lambda_k| \leq 2T_n} \frac{a_k}{z + \lambda_k} \right| = O\left(\sum_{|\lambda_k| \leq 2T_n} \frac{|a_k|}{|\text{Re}(\lambda_k) + \text{Re}(z)|} \right)$$

$$(9) \qquad = o\left(\sum_{|\lambda_k| \leq 2T_n} \frac{|\lambda_k|^{\sigma_0}}{||\text{Re}(\lambda_k)| - |\text{Re}(z)||} \right).$$

Let $\{\lambda_{n_k}\}$ be the sequence determined by Lemma 4.5(a), and set

$$T_k = \frac{1}{2} \left(\text{Re}(\lambda_{n_k}) + \text{Re}(\lambda_{n_k+1}) \right).$$

For any fixed positive number h and any number T, there is a uniform bound on the number of elements of the sequence $\{r_k\}$ in the interval $[T, T+h]$. Also, there is a positive constant c such that $|r_k - T_n| \geq c$ for all k and n. With these facts and Lemma 4.2, (9) can be bounded by

$$o\left(\sum_{k=1}^{T_n^{\sigma_1} - cT_n^{\sigma_1 - 1}} \frac{k^{\sigma_0/\sigma_1}}{T_n - k^{1/\sigma_1}} \right),$$

which we can estimate by the integral

$$o\left(\int_0^{T_n^{\sigma_1} - cT_n^{\sigma_1 - 1}} \frac{x^{\sigma_0/\sigma_1}}{T_n - x^{1/\sigma_1}} dx \right) = o\left(\int_0^{T_n - c} \frac{u^{\sigma_0}}{T_n - u} d\left[u^{\sigma_1} \right] \right)$$

$$= o\left(T_n^{\sigma_0 + \sigma_1 - 1} \log T_n \right).$$

Similarly, for part (b), we can bound the finite sum in the spectral case by the integral

$$O\left(\int_0^{T_n - c} \frac{1}{T_n - x} d\left[x^\alpha (\log x)^{-\beta} \right] \right) = O\left(T_n^{\alpha - 1} (\log T_n)^{-\beta + 1} \right)$$

$$= O\left(T_n^{-\text{Re}(p_0) - 1} (\log T_n)^{m(p_0) + 1} \right).$$

With this, the proof of Proposition 4.6, and Theorem 4.1, is complete. $\square$

§5. Asymptotics in a parallel strip.

In this section we will consider a regularized harmonic series $R(z)$ corresponding to a sequence $\{\lambda_k\}$ that converges to infinity in a strip, meaning there exists a constant C such that for all k, we have the inequality

$$|\text{Im}(\lambda_k)| < C.$$

Recall that the sequence $\{\lambda_k\}$ is such that $\text{Re}(\lambda_k) \to \infty$. An individual term $1/(z+\lambda_k)$ may become arbitrarily large if z approaches λ_k. Under the assumption $|\text{Im}(\lambda_k)| < C$, we will show that by bounding z to a parallel horizontal strip given by

$$0 < c < |\text{Im}(z) - \text{Im}(\lambda_k)| < c' < \infty,$$

we shall be able to determine the asymptotic behavior of the regularized harmonic series $R(z)$.

Theorem 5.1. *Assume there is a constant C such that for all k we have $|\text{Im}(\lambda_k)| \leq C$. Let $z = -T + iy$ and assume there are constants $c, c' > 0$ such that for all k, we have*

$$0 < c < |\text{Im}(z) - \text{Im}(\lambda_k)| < c' < \infty$$

a) *In the notation of **DIR 2**, we have*

$$R(-T + iy) = o(T^{\sigma_1 + \sigma_0 - 1} \log T)$$

uniformly as $T \to \infty$.

b) *Let $m = m(p_0) + 1$ if there is a complex index p with $\text{Re}(p_0) = \text{Re}(p) \in \mathbf{Z}_{<0}$ and $b_p \neq 0$, otherwise simply set $m = m(p_0)$. Then for any q with $\text{Re}(q) > 0$, we have*

$$R(-T + iy) = O(T^{-\text{Re}(p_0) - 1}(\log T)^m)$$

uniformly as $T \to \infty$. Since $-\text{Re}(p_0) - 1 \leq M < -\text{Re}(p_0)$, we have

$$R(-T + iy) = O(T^M (\log T)^m)$$

uniformly as $T \to \infty$.

Proof. The proof follows that of Theorem 5.1. The only aspect of the proof which needs to be reconsidered is the asymptotic behavior of the finite sum

$$\sum_{|\lambda_k| \leq 2T} \frac{a_k}{u + w + \lambda_k}.$$

However, we immediately have the estimate

$$\left| \sum_{|\lambda_k| \leq 2T} \frac{a_k}{z + \lambda_k} \right| = O\left(\sum_{|\lambda_k| \leq 2T} \frac{|a_k|}{c + |\mathrm{Re}(\lambda_k) + \mathrm{Re}(z)|} \right)$$

$$= o\left(\sum_{|\lambda_k| \leq 2T} \frac{|\lambda_k|^{\sigma_0}}{c + ||\mathrm{Re}(\lambda_k)| - |\mathrm{Re}(z)||} \right),$$

which can be bounded by the sum

$$o\left(\sum_{k=1}^{c_1 T^{\sigma_1}} \frac{k^{\sigma_0/\sigma_1}}{c + |c_2 k^{1/\sigma_1} - T|} \right).$$

As in the proof of Proposition 4.6, this sum can be estimated by the integral

$$o\left(\int_0^{c_1 T^{\sigma_1}} \frac{x^{\sigma_0/\sigma_1}}{|T - x^{1/\sigma_1}| + c} dx \right) = o\left(\int_0^{c_1' T} \frac{u^{\sigma_0}}{|T - u| + c} d[u^{\sigma_1}] \right)$$

$$= o\left(T^{\sigma_0 + \sigma_1 - 1} \log T \right).$$

The proof of the remaining parts of the theorem follows the same pattern. $\square$

§6. Regularized product and series type.

In §1 we defined what is meant for a function to be a regularized product. In this section, we define functions which are of regularized product type. After this, we will summarize the asymptotic formulae of the previous sections for this new class of functions.

Definition 6.1. A meromorphic function of finite order $G(z)$ is said to be of **regularized product type** if

$$G(z) = Q(z)e^{P(z)} \prod_{j=1}^{n} D_j(\alpha_j z + \beta_j)^{k_j}$$

where:
  i) $Q(z)$ is a rational function;
 ii) $P(z)$ is a polynomial;
iii) $D_j(z)$ is a regularized product, with $\alpha_j, \beta_j \in \mathbf{C}$ and $k_j \in \mathbf{Z}$;
 iv) the α_j, β_j are restricted so that the zeros and poles of $D_j(\alpha_j z + \beta_j)$ lie in the the union of regions of the form:
   • a sector opening to the right, meaning

$$\{z \in \mathbf{C} : -\frac{\pi}{2} + \epsilon < \arg(z) < \frac{\pi}{2} - \epsilon\} \text{ for some } \epsilon > 0,$$

   • a sector opening to the left, meaning

$$\{z \in \mathbf{C} : \frac{\pi}{2} + \epsilon < \arg(z) < \frac{3\pi}{2} - \epsilon\} \text{ for some } \epsilon > 0,$$

   • a vertical strip, meaning

$$\{z \in \mathbf{C} : a < \mathrm{Re}(z) < b\} \text{ for some } a, b \in \mathbf{R}.$$

If G is of regularized product type, then any vertical strip which contains at most a finite number of zeros of G is called an **admissible strip**. Observe that if the zeros and poles of the function G are in sectors, as described in the first two conditions of Definition 6.1(iv) above, then every vertical strip is admissible. The main results of [JoL 93c] will be assumed (see in particular Theorem 1.5), and will not be repeated in these notes. Many examples of functions of regularized product type are given in §7 of [JoL 93c].

Corresponding to the multiplicative group of regularized product types, we have the additive group of their logarithmic derivatives, namely

$$G'/G(s) = \alpha_0 D_0'/D_0(\alpha_0 z + \beta_0) + \sum_{j=1}^{n} k_j \alpha_j D_j'/D_j(\alpha_j z + \beta_j).$$

This leads us to define the additive notion for its own sake.

A function R will be said to be of **regularized harmonic series type** if it is a finite linear combination

$$R(z) = \sum c_j R_j(\alpha_j z + \beta_j) + P'(z) + \sum \frac{c_k'}{z - \beta_k'},$$

where $c_j, c_k', \alpha_j, \beta_j, \beta_k' \in \mathbf{C}$, each R_j is a regularized harmonic series as defined in §1, P' is a polynomial, and α_j and β_j are subject to condition (iv) with respect to the poles of $R_j(\alpha_j z + \beta_j)$. Sections §2 to §5 provide systematic estimates for functions of regularized harmonic series type in vertical and horizontal strips, not just for those which happen to be logarithmic derivatives of regularized product types. Indeed, a transformation $z \mapsto \alpha z + \beta$ amounts to a dilation, rotation and translation, so the estimates of §2 and §5 apply to each term in the above linear combination for a function of regularized harmonic series type. The resulting estimates will be summarized in Theorem 6.2 below.

Let $D_0(z) = Q(z)e^{P(z)}$ and define the **reduced order** of D_0 to be $(M_0, 0)$ where $M_0 = \max\{0, \deg P - 1\}$. Suppose G is of regularized product type of the form

$$G(z) = D_0(\alpha_0 z + \beta_0) \prod_{j=1}^{n} D_j(\alpha_j z + \beta_j)^{k_j}.$$

Let (M_j, m_j) be the reduced order of D_j, as defined at the end of Chapter I, §1. then we define the **reduced order** of G to be the pair of integers (M, m) where $M = \max\{M_j\}$ and m is the maximum over all m_j for which $M = M_j$.

We make a similar definition for the reduced order (M, m) of a regularized harmonic series type. In this case, the reduced order of P' is $(\deg P', 0)$ if $P' \neq 0$.

We shall use for a vertical strip the notation

$$\text{Str}(x_1, x_2) = \{z \in \mathbf{C} : x_1 \leq \text{Re}(z) \leq x_2\}.$$

The asymptotic formulas of the previous sections can now be easily summarized for any function R of regularized harmonic series type of reduced order (M, m).

Theorem 6.2. *Let R be of regularized harmonic series type of reduced order (M, m).*

a) *Let $S = \text{Str}(x_1, x_2)$ be an admissible strip. Then uniformly for $x \in [x_1, x_2]$, we have the asymptotic relation*

$$R(x + iy) = O(y^M (\log y)^{m+1}) \quad \text{as } |y| \to \infty.$$

b) *Let $S = \text{Str}(x_1, x_2)$ be a vertical strip which contains an infinite number of poles of the function R. Then there is a sequence of real numbers $T_n \to \infty$ such that for all $x \in [x_1, x_2]$, we have the following uniform asymptotic relation*

$$R(x \pm iT_n) = O(T_n^M (\log T_n)^{m+1}) \quad \text{as } T_n \to \infty.$$

Proof. Assertion (a) follows directly from the definition of regularized harmonic series type, admissible strip, and Corollary 2.2, Corollary 3.2, and Theorem 5.1. Assertion (b) follows from the same results together with Theorem 4.1. Note that in order to obtain the symmetry as asserted, one needs to consider the union of the sequences in the regularized harmonic series decomposition of G, which a possible change of sign. $\square$

§7. Some examples.

To conclude this chapter, we shall briefly discuss how the results of the previous sections contain various classical asymptotic formula for the gamma function and the Riemann zeta function, and improve an existing bound for the Selberg zeta function associated to a compact hyperbolic Riemann surface.

Example 1: The gamma function. The classical gamma function $\Gamma(s)$ is, up to a factor of the form e^{as+b}, the regularized product associated to the sequence $\mathbf{Z}_{\geq 0}$. In the notation used above, we have $-\text{Re}(p_0) = 1$ and $m(p_0) = 0$, hence the gamma function is of regularized product type of reduced order $(0,0)$. Theorem 2.3 yields the classical Stirling's formula for $\log\Gamma(s)$. As for asymptotics in vertical strips, Corollary 2.2 yields the equally classical asymptotic formula

$$\Gamma'/\Gamma(s) = \log|s| + O(1)$$

as $s \to \infty$ in any vertical strip of finite width. This result is an important ingredient in the proof of the classical explicit formula for zeta functions of number fields (see, for example, Chapter XVII of [La 70]).

Example 2: Dirichlet polynomials. We define a **Dirichlet polynomial** to be a holomorphic function of the form

$$P(s) = \sum_{n=1}^{N} a_n b_n^s$$

where $\{a_n\}$ is a finite sequence of complex numbers, $\{b_n\}$ is a finite sequence of positive real numbers. In Chapter II we will apply the general Cramér theorem from [JoL 93c] to show that P is of regularized product type of reduced order $(0,1)$.

Example 3: The Riemann zeta function. Let $\zeta_{\mathbf{Q}}(s)$ be the Riemann zeta function and consider the sequences

$$\Lambda_+ = \{\rho/i \in \mathbf{C} \mid \zeta_{\mathbf{Q}}(\rho) = 0, \ 0 \leq \text{Re}(\rho) \leq 1, \ \text{Im}(\rho) > 0\}$$

and

$$\Lambda_- = \{\rho/(-i) \in \mathbf{C} \mid \zeta_{\mathbf{Q}}(\rho) = 0, \ 0 \leq \text{Re}(\rho) \leq 1, \ \text{Im}(\rho) < 0\}.$$

By Corollary 1.3 of [JoL 93c], the theta function θ_{Λ_+} associated to the sequence Λ_+ satisfies the asymptotic conditions **AS 1**, **AS 2**, and **AS 3** with $-\mathrm{Re}(p_0) = 1$ and $m(p_0) = 1$. Similarly, the theta function θ_{Λ_-} associated to the sequence Λ_- satisfies the asymptotic conditions **AS 1**, **AS 2**, and **AS 3**, again with $-\mathrm{Re}(p_0) = 1$ and $m(p_0) = 1$. Theorem 1.6 of [JoL 93a] then implies that the functions

$$D_+(z) = \exp\left(-\mathrm{CT}_{s=0}\mathbf{LM}\theta_{\Lambda_+}(s, z)\right)$$

and

$$D_-(z) = \exp\left(-\mathrm{CT}_{s=0}\mathbf{LM}\theta_{\Lambda_-}(s, z)\right)$$

are holomorphic functions of finite order with

$$D_+(z) = 0 \quad \text{if and only if } -z \in \Lambda_+,$$

and

$$D_-(z) = 0 \quad \text{if and only if } -z \in \Lambda_-.$$

The argument of §6 of [JoL 93c] produces the relation

$$\zeta_{\mathbf{Q}}(s)\Gamma(s/2) = D_+(-s/i)D_-(s/i)(s(s-1))^{-1}e^{as+b}$$

for some constants a and b. Therefore, the Riemann zeta function is of regularized product type of reduced order $(0, 1)$.

Example 3 and Theorem 5.1 combine to assert the existence of a sequence of numbers $\{T_n\}$ with $T_n \to \infty$ such that

$$\left|\zeta_{\mathbf{Q}}'/\zeta_{\mathbf{Q}}(x \pm iT_n)\right| = O\left((\log T_n)^2\right)$$

for $x \in [x_1, x_2]$ and $T_n \to \infty$. This result is proved classically in a different manner, see, for instance [In 32], pages 71-73.

Example 4: The Selberg zeta function $Z(s)$ associated to a compact Riemann surface. The results of [JoL 93c] show that the Selberg zeta function $Z(s)$ associated to any compact hyperbolic Riemann surface is an entire function of regularized product type and reduced order $(1, 1)$, since, in this case, $-\mathrm{Re}(p_0) = 2$ and $m(p_0) = 1$. Theorem 5.1 asserts the existence of a sequence of numbers $\{T_n\}$ with $T_n \to \infty$ such that

$$|Z'/Z(x \pm iT_n)| = O\left(T_n(\log T_n)^2\right)$$

for $x \in [x_1, x_2]$ and $T_n \to \infty$. This improves a result stated on page 80 of [He 76], giving an upper bound of the form $O(T_n^2)$. As we will see in subsequent chapters, this improvement is significant since it allows us to apply the explicit formula for the Selberg zeta function to any test function that satisfies the three basic conditions to order one, which essentially amounts to requiring that the test function and its first derivative satisfies certain smoothness conditions. With the weaker bound from page 80 of [He 76], one would be forced to require smoothness conditions on a test function and its first two derivatives.

Example 5: The Selberg zeta function $Z(s)$ associated to a non-compact Riemann surface. One can apply the results of this section to the Selberg zeta function associated to any non-compact hyperbolic Riemann surface of finite volume X since, by Theorem 7.1 of [JoL 93c], this Selberg zeta function is of regularized product type of reduced order $(1,1)$. Associated to X there is a meromorphic function called the scattering determinant, which is the determinant of the constant terms in the Fourier expansions in the cusps of the Eisenstein series (see the discussion starting on page 498 of [He 83] or page 49 of [Sel 56]). Theorem 7.1 of [JoL 93c] shows that the scattering determinant is of regularized product type of reduced order $(0,1)$.

In [JoL 94] we will give a more complete discussion of the application of our results to scattering determinants and to Eisenstein series on non-compact hyperbolic Riemann surfaces of finite volume.

A larger but non-exhaustive list of further examples is given at the end of [JoL 93c], including zeta functions arising in representation theory and the theory of modular forms, and zeta functions associated with certain higher dimensional manifolds. We shall return to these specific applications in a subsequent publication devoted exclusively to them.

CHAPTER II

Cramér's theorem as an explicit formula.

In [Cr 19] Cramér showed that if $\{\rho_k\}$ ranges over the non-trivial zeros of the Riemann zeta function with $\text{Im}(\rho_k) > 0$, then the series

$$V(z) = \sum e^{\rho_k z}$$

converges for $\text{Im}(z) > 0$ and has a singularity at the origin of the type $\log z/(1 - e^{-z})$, by which we mean that the function

$$F(z) = 2\pi i V(z) - \frac{\log z}{1 - e^{-z}},$$

which is defined for $\text{Im}(z) > 0$, has a meromorphic continuation to all $\mathbf{C}$, with simple poles at the points $\pm \pi i n$, where n ranges over the integers, and at the points $\pm \log p^m$, where p^m ranges over the prime powers. In [JoL 93c] we proved an analogous theorem for any meromorphic function with an Euler sum and functional equation whose fudge factors which are of regularized product type. In this chapter we will prove a Cramér-type theorem by considering the same contour integral as analyzed in [JoL 93c] for a more general class of test functions. Specifically, we consider the contour integral

$$\frac{1}{2\pi i} \int_{\mathcal{R}_a} \phi(s) Z'/Z(s) ds$$

over a semi-infinite vertical rectangle $\mathcal{R}_a$ which is assumed to contain the top half of the "critical strip" of Z. The test function ϕ is assumed to be holomorphic on the closure of $\mathcal{R}_a$ and to have reasonably weak growth conditions, essentially what is needed to make the proof go through. In the Cramér theorem from [JoL 93c], the test function ϕ depended on a complex parameter z, namely

$$\phi(s) = \phi_z(s) = e^{sz}.$$

For $\mathrm{Im}(z) > 0$, the function ϕ_z has exponential decay when s lies in a finite strip and $\mathrm{Im}(s) \to \infty$. As a result, the analysis needed in [JoL 93c] required a very weak growth result, which we proved for general meromorphic functions of prescribed order (see, in particular, Lemma 2.1 of [JoL 93c]).

In §1 we recall the fundamental class of functions which we have defined, and discuss its relation to the Selberg class of functions defined in [Sel 91]. The definitions of §1 are used throughout, but the subsequent sections may be logically omitted for the rest of this work. Taking into account the asymptotic estimates of Chapter I, our method of proof from [JoL 93c] applies to the more general class of test functions considered here. In §2 we establish notation and state the main result of this chapter. The proof is given in §3, and various applications are discussed in §4. As remarked on page 397 of [JoL 93c], our proof does differ from the original proof due to Cramér, which is one of the reasons why we can easily generalize the theorems to the class of functions which have Euler sums and functional equations with fudge factors which are of regularized product type.

§1. Euler sums and functional equations.

We shall say that the functions Z and $\tilde{Z}$ have an **Euler sum** and **functional equation** if the following properties are satisfied:

1. **Meromorphy.** The functions Z and $\tilde{Z}$ are meromorphic functions of finite order.

2. **Euler Sum.** There are sequences $\{\mathbf{q}\}$ and $\{\tilde{\mathbf{q}}\}$ of real numbers > 1 that depend on Z and $\tilde{Z}$, respectively, and that converge to infinity, such that for every $\mathbf{q}$ and $\tilde{\mathbf{q}}$, there exist complex numbers $c(\mathbf{q})$ and $c(\tilde{\mathbf{q}})$ and $\sigma_0' \geq 0$ such that for all $\mathrm{Re}(s) > \sigma_0'$,

$$\log Z(s) = \sum_{\mathbf{q}} \frac{c(\mathbf{q})}{\mathbf{q}^s} \quad \text{and} \quad \log \tilde{Z}(s) = \sum_{\tilde{\mathbf{q}}} \frac{c(\tilde{\mathbf{q}})}{\tilde{\mathbf{q}}^s}.$$

The series are assumed to converge uniformly and absolutely in any half-plane of the form $\mathrm{Re}(s) \geq \sigma_0' + \epsilon > \sigma_0'$.

3. **Functional Equation.** There are functions G and $\tilde{G}$, meromorphic and of finite order, and there exists σ_0 with $0 \leq \sigma_0 \leq \sigma_0'$ such that

$$Z(s)G(s) = \tilde{Z}(\sigma_0 - s)\tilde{G}(\sigma_0 - s).$$

We let

$$\Phi(s) = G(s)/\tilde{G}(\sigma_0 - s),$$

so the functional equation can be written in the form

$$Z(s)\Phi(s) = \tilde{Z}(\sigma_0 - s).$$

We call G, $\tilde{G}$ or Φ the **fudge factors** of the functional equation.

Remark 1. We inadvertently took $\sigma_0 = \sigma_0'$ in [JoL 93c], but it is important to allow $\sigma_0 \neq \sigma_0'$ for some applications, e.g. the scattering determinants which we considered in §7 of [JoL 93c]. No change is needed in the proof of the Cramér theorem except for choosing $a > 0$ such that $\sigma_0 + a > \sigma_0'$ as in §2 below. We dealt with the scattering determinant in [JoL 93c], in the context of Cramér's theorem.

The Euler sum for Z implies that Z is uniformly bounded for $\mathrm{Re}(s) \geq \sigma_0' + \epsilon$ for every $\epsilon > 0$. Notice that Z has no zeros or poles for $\mathrm{Re}(s) > \sigma_0'$, and all zeros and poles of Z in the region $\mathrm{Re}(s) < -a$ agree in location and order with poles and zeros of Φ.

Remark 2. A Dirichlet series expression is assumed only for Z and $\tilde{Z}$, so the fudge factors do not occur symetrically for the zeta function in the above conditions, although they might appear to do so in the functional equation. For example, in the most classical case of the Riemann zeta function, the fudge factor is essentially the gamma function, which does not have a Dirichlet series expansion but is of regularized product type.

We define a triple $(Z, \tilde{Z}, \Phi)$ to be in **the fundamental class** if Z, $\tilde{Z}$ have an Euler sum and functional equation, and the fudge factors are of regularized product type. Selberg has defined a "Selberg class" of functions in analytic number theory (see [Se 91], [CoG 93]). Our class is much wider than Selberg's class in several major respects:

1) Selberg's fudge factors are of gamma type, i.e. $\Gamma(\alpha s + \beta)$.
2) Selberg assumes a Ramanujan-Petersson estimate on the coefficients of the Euler sum, but we do not.
3) Selberg's Euler sum involves ordinary integers and ordinary prime powers. We allow arbitrary positive numbers $\{\mathbf{q}\}$.

Our conditions allow for a much wider domain of applicability in spectral theory as illustrated by our varied examples, and many more to be treated in subsequent papers. For example, our conditions allow the fudge factors to include Γ_2, Γ_d for general d, or for ζ and L functions themselves, or any function of regularized product type.

Remark 3. Even so, the Euler sum condition is still not sufficiently general for our purposes and will be ultimately be generalized to a Bessel sum condition. For further comments on this point of view, see §3 and §4 of Chapter V. In the same vein, the functional equation will also be replaced by an additive relation where the additive fudge factor will be assumed to be a regularized harmonic series, or regularized harmonic series type, in analogy with regularized product type.

§2. The general Cramér formula.

Let $a > 0$ be such that $\sigma_0 + a > \sigma_0'$. We define the following regions in the complex plane:

$$\mathcal{R}_a^+ = \text{semi-infinite open rectangle bounded by the lines}$$

$$\text{Re}(s) = -a, \quad \text{Re}(s) = \sigma_0 + a, \quad \text{Im}(s) = 0.$$

$$\mathcal{R}_a^+(T) = \text{the portion of } \mathcal{R}_a^+ \text{ below the line } \text{Im}(s) = T.$$

We allow Z and Φ to have zeros or poles on the finite real segment $[-a, \sigma_0 + a]$, but we assume that Φ and Z have no zeros or poles on the vertical edges with $\text{Re}(s) = -a$ and $\text{Re}(s) = \sigma_0 + a$ and $\text{Im}(s) > 0$.

Let ϕ be any function which is holomorphic on the closure of the semi-infinite rectangle $\bar{\mathcal{R}}_a^+$, and let H be a meromorphic function on this closure. We are interested in studying the (formal) sums

$$(1) \qquad \text{div}_{H,a}^+(\phi) = \sum_{z \in \mathcal{R}_a^+} v_H(z)\phi(z)$$

where

$$v_H(z) = \text{ord}_H(z)$$

is the order of the zero or pole of H at z, so the sum (1) is actually over the divisor of H which lies in $\mathcal{R}_a^+$. Such sums do not converge a priori, so we need to define them as limits in a suitable sense, and for suitable functions ϕ. On a space of functions decaying sufficiently fast (depending on H), the divisor

$$\text{div}_{H,a}^+ = \sum_{z \in \mathcal{R}_a^+} v_H(z)(z)$$

gives rise to the functional defined by the sum (1). The functional itself may be denoted by $[\text{div}_{H,z}^+]$ to distinguish the functional from the divisor. To determine such a space of functions, we proceed as follows.

Let $H_1, \ldots, H_r$ be a finite number of functions which are meromorphic on the closure of $\mathcal{R}_a^+$, and let ϕ be a function which is holomorphic on this closure. We say that a sequence $\{T_m\}$ of positive real numbers tending to infinity is ϕ–**admissible** for $\{H_1, \ldots, H_r\}$ if for any k, H_k has no zero or pole on the segment

$$S_m = [-a + iT_m, \sigma_0 + a + iT_m]$$

and

$$\phi(s)H'_k/H_k(s) \to 0 \quad \text{for } s \in S_m \text{ with } m \to \infty.$$

When $\{H_1, \ldots, H_r\}$ is the set of functions $\{Z, \tilde{Z}, \Phi\}$, we say simply that $\{T_m\}$ is **admissible**. With respect to a ϕ–admissible sequence $\{T_m\}$, we define the **divisor functional for H**, which we denote by $\mathrm{div}_H^+(\phi)$, to be the limit

$$(2) \qquad \mathrm{div}_{H,a}^+(\phi) = \lim_{m \to \infty} \sum_{z \in \mathcal{R}_a^+(T_m)} v_H(z)\phi(z),$$

if such a limit exists. In particular, for $H = Z$, we let

$$\{\rho\} = \text{set of zeros and poles of } Z \text{ in } \mathcal{R}_a^+$$

so the sum (2) can be written as

$$\mathrm{div}_{Z,a}^+(\phi) = \lim_{m \to \infty} \sum_{\rho \in \mathcal{R}_a^+(T_m)} v_Z(\rho)\phi(\rho).$$

We define other functionals as follows. Here we do not try to give subtle conditions on what amounts to a half-Fourier transform, so we simply assume that the derivative $\phi'(s)$ is in L^1 of each vertical half line $[-a, -a + i\infty]$ and $[\sigma_0 + a, \sigma_0 + a + i\infty]$. We define the **positive Cramér functional for q**, which we denote by $\mathrm{Cr}_{q,a}^+(\phi)$, to be the integral

$$\mathrm{Cr}_{q,a}^+(\phi) = \int_{\sigma_0+a}^{\sigma_0+a+i\infty} \phi'(s)q^{-s}ds.$$

Similarly, the **negative Cramér functional for q** is defined by the integral

$$\mathrm{Cr}_{q,a}^-(\phi) = \int_{\sigma_0+a-i\infty}^{\sigma_0+a} \phi'(\sigma_0 - s)q^{-s}ds.$$

Also, with respect to an H-admissible sequence $\{T_m\}$, where H is a meromorphic function which is holomorphic on $\mathrm{Re}(s) = -a$, we define the functional

$$U^+_{H,-a}(\phi) = \lim_{m \to \infty} \int\limits_{-a}^{-a+iT_m} \phi(s)H'/H(s)ds.$$

When all the above functionals are defined, we then consider

The Cramér formula.

$$2\pi i \mathrm{div}^+_{Z,a}(\phi) = \sum_{\tilde{\mathbf{q}}} c(\tilde{\mathbf{q}})\mathrm{Cr}^-_{\tilde{\mathbf{q}},a}(\phi) - \sum_{\mathbf{q}} c(\mathbf{q})\mathrm{Cr}^+_{\mathbf{q},a}(\phi)$$

$$+ U^+_{\Phi,-a}(\phi) + \int\limits_{-a}^{\sigma_0+a} \phi'(s)\log Z(s)ds$$

$$+ \phi(-a)\left(\log \tilde{Z}(\sigma_0 + a) - \log Z(-a)\right).$$

Such a formula is derived formally by considering the contour integral

$$\frac{1}{2\pi i} \int\limits_{\partial \mathcal{R}^+_a} \phi(s)Z'/Z(s)ds,$$

which can be evaluated in one way by using the residue theorem, and in another way by using the Euler sum and functional equation for Z.

We are interested in conditions on ϕ for which the Cramér formula holds. For this purpose, we consider the following growth assumptions on ϕ.

GR 1. $\phi(s)\Phi'/\Phi(s)$ is in L^1 on the vertical ray $[-a, -a + i\infty]$.

GR 2. The derivative $\phi'(s)$ is in L^1 of each vertical half line

$$[-a, -a+i\infty] \quad \text{and} \quad [\sigma_0 + a, \sigma_0 + a + i\infty].$$

The first condition compares the decay of $\phi(s)$ with the exponential growth of $\Phi(s)$. As in [JoL 93c], one often considers the situation when Φ is of regularized product type. In this case, the above growth conditions can be verified as follows.

Proposition 2.1.

a) *If Φ is of regularized product type of reduced order (M, m) and the ray $[-a, -a+i\infty]$ lies in an admissible strip (defined in Chapter I, §6), then*

$$\Phi'/\Phi(s) = O(|s|^M (\log |s|)^m)$$

for $|s| \to \infty$ and s on the ray $[-a, -a + i\infty]$.
b) *If for some $\delta > 0$, ϕ has the decay*

$$\phi(s) = O(1/|s|^{M+1}(\log |s|)^{m+1+\delta}) \quad \text{for } |s| \to \infty \text{ on the ray,}$$

then **GR 1** *is satisfied.*

The hypotheses in the above criterion have been shown to be satisfied in several cases which are of direct interest in our theory of regularized products, for instance Theorem 5.2 of Chapter I. The next theorem asserts that the divisor functional $[\operatorname{div}^+_{Z,a}]$ is defined on the vector space of functions satisfying **GR 1** and **GR 2**, and satisfies the Cramér formula.

Theorem 2.2. *Assume ϕ and Φ satisfy the two growth conditions* **GR 1** *and* **GR 2**. *Then all the functionals $\operatorname{div}^+_{Z,a}(\phi)$, $U^+_{\Phi,-a}(\phi)$, $\operatorname{Cr}^+_{q,a}(\phi)$, and $\operatorname{Cr}^-_{q,a}(\phi)$ are defined, and the following formula holds:*

$$2\pi i \operatorname{div}^+_{Z,a}(\phi) = \sum_{\tilde{q}} c(\tilde{q}) \operatorname{Cr}^-_{\tilde{q},a}(\phi) - \sum_{q} c(q) \operatorname{Cr}^+_{q,a}(\phi)$$

$$+ U^+_{\Phi,-a}(\phi) + \int_{-a}^{\sigma_0+a} \phi'(s) \log Z(s) ds$$

$$+ \phi(-a) \left(\log \tilde{Z}(\sigma_0 + a) - \log Z(-a) \right).$$

As stated above, the proof of Theorem 2.2 will be given in the following section, and various applications of the theorem will be discussed in §4.

§3. Proof of the Cramér theorem.

The pattern of proof of Theorem 2.2 follows §2 of [JoL 93c], which, as we shall remark below, contains one significant technical improvement over the proof of the original theorem given by Cramér [Cr 19] for the Riemann zeta function.

Choose an $\epsilon > 0$ sufficiently small so that Z has no zeros or poles in the open rectangle with vertices

$$-a, \quad -a+i\epsilon, \quad \sigma_0+a+i\epsilon, \quad \sigma_0+a$$

or on the line segment $[-a+i\epsilon, \sigma_0+a+i\epsilon]$. Note that the function Z may have zeros or poles on the horizontal line segment $[-a, \sigma_0+a]$. For T sufficiently large, we shall study the contour integral

$$2\pi i V_Z(z,\epsilon;T) = \int_{-a+iT}^{-a+i\epsilon} \phi(s)Z'/Z(s)ds + \int_{-a+i\epsilon}^{\sigma_0+a+i\epsilon} \phi(s)Z'/Z(s)ds$$

$$(1) \qquad + \int_{\sigma_0+a+i\epsilon}^{\sigma_0+a+iT} \phi(s)Z'/Z(s)ds \int_{\sigma_0+a+iT}^{-a+iT} \phi(s)Z'/Z(s)ds.$$

We may assume that Z has no zeros or poles on the line segment connecting the points $-a+iT$ and σ_0+a+iT, because we will pick $T = T_m$ for m sufficiently large. Let:

$\mathcal{R}_T(\epsilon) =$ the finite rectangle with vertices

$$-a+iT, \quad -a+i\epsilon, \quad \sigma_0+a+i\epsilon, \quad \sigma_0+a+iT.$$

By the residue theorem, we have

$$2\pi i V_Z(z,\epsilon;T) = \sum_{\rho \in \mathcal{R}_T(\epsilon)} v(\rho)\phi(\rho).$$

Theorem 2.2 will be established by studying each of the four integrals in (1). For simplicity, let us call these integrals the left, bottom, right, and top integrals, respectively. We begin with the top integral which will be shown to be arbitrarily small upon letting $T = T_m$ approach infinity.

Lemma 3.1. *Let* $\{T_m\}$ *be an admissible sequence relative to* Z. *Then we have*

$$
\lim_{m \to \infty} \left[\int_{\sigma_0+a+iT_m}^{-a+iT_m} \phi(s)Z'/Z(s)ds \right] = 0.
$$

The proof of Lemma 3.1 follows directly from the definition of an admissible sequence.

To continue, we have, from the growth assumption **GR 1**, the limit

$$
\lim_{m \to \infty} \int_{-a+iT_m}^{-a+i\epsilon} \phi(s)Z'/Z(s)ds = \int_{-a+i\infty}^{-a+i\epsilon} \phi(s)Z'/Z(s)ds
$$

and

$$
\lim_{m \to \infty} \int_{\sigma_0+a+i\epsilon}^{\sigma_0+a+iT_m} \phi(s)Z'/Z(s)ds = \int_{\sigma_0+a+i\epsilon}^{\sigma_0+a+i\infty} \phi(s)Z'/Z(s)ds.
$$

By combining these equations with Lemma 3.1, we have the following preliminary result.

Proposition 3.2. *With notation as above, we have*

$$
2\pi i V_Z(z, \epsilon) = \lim_{m \to \infty} 2\pi i V_Z(z, \epsilon; T_m)
$$

$$
= \int_{-a+i\infty}^{-a+i\epsilon} \phi(s)Z'/Z(s)ds + \int_{-a+i\epsilon}^{\sigma_0+a+i\epsilon} \phi(s)Z'/Z(s)ds
$$

$$
+ \int_{\sigma_0+a+i\epsilon}^{\sigma_0+a+i\infty} \phi(s)Z'/Z(s)ds.
$$

As before, let us call the integrals in Proposition 3.2 the left, bottom and right integrals, respectively. By the above stated assumption on ϵ, we have

$$
V_Z(z, \epsilon) = V_Z(z).
$$

To continue our proof of Theorem 2.2, we will compute the three integrals in Proposition 3.2 using the axioms of Euler sum and functional equation. After these computations, we will let ϵ approach 0, which will complete the proof.

Let us use the functional equation to re-write the left integral as the sum of three integrals involving $\tilde{Z}$ and Φ. Specifically, we have

$$\int_{-a+i\infty}^{-a+i\epsilon} \phi(s)Z'/Z(s)ds = \int_{-a+i\infty}^{-a+i\epsilon} \phi(s)\left[-\Phi'/\Phi(s) - \tilde{Z}'/\tilde{Z}(\sigma_0 - s)\right]ds$$

$$(2) \qquad\qquad = \int_{-a+i\epsilon}^{-a+i\infty} \phi(s)\Phi'/\Phi(s)ds$$

$$(3) \qquad\qquad + \int_{\sigma_0+a-i\infty}^{\sigma_0+a-i\epsilon} \phi(\sigma_0 - s)\tilde{Z}'/\tilde{Z}(s)ds.$$

After we let $\epsilon \to 0$, the integral in (2) appears in the statement of Theorem 2.2 as the functional $U_\Phi^+(\phi)$. Note that letting $\epsilon \to 0$ is justified since Φ was assumed to be holomorphic and non-zero on the vertical lines of integration. As for (3), we can re-write this integral using the Euler sum of $\tilde{Z}$, yielding

$$\int_{\sigma_0+a-i\infty}^{\sigma_0+a-i\epsilon} \phi(\sigma_0 - s)\tilde{Z}'/\tilde{Z}(s)ds$$

$$= \phi(\sigma_0 - s)\log\tilde{Z}(s)\Big|_{\sigma_0+a-i\infty}^{\sigma_0+a-i\epsilon} + \int_{\sigma_0+a-i\infty}^{\sigma_0+a-i\epsilon} \phi'(\sigma_0 - s)\log\tilde{Z}(s)ds$$

$$= \phi(-a + i\epsilon)\log\tilde{Z}(\sigma_0 + a - i\epsilon)$$

$$+ \sum_{\tilde{q}} c(\tilde{q}) \int_{\sigma_0+a-i\infty}^{\sigma_0+a-i\epsilon} \phi'(\sigma_0 - s)\tilde{q}^{-s}ds.$$

By the Euler sum condition and the fact that $a > 0$, we can let ϵ

approach zero to get the equality

$$\int_{\sigma_0+a-i\infty}^{\sigma_0+a} \phi(\sigma_0 - s)\tilde{Z}'/\tilde{Z}(s)ds$$

(4)
$$= \phi(-a)\log \tilde{Z}(\sigma_0 + a) + \sum_{\tilde{q}} c(\tilde{q})\mathrm{Cr}_{\tilde{q},a}^{-}(\phi).$$

Both terms in (4) appear in the statement of Theorem 2.2.

In the same manner as above, the right integral can be re-written using the Euler sum of Z, yielding

$$\int_{\sigma_0+a+i\epsilon}^{\sigma_0+a+i\infty} \phi(s)Z'/Z(s)ds$$

$$= \phi(s)\log Z(s)\Big|_{\sigma_0+a+i\epsilon}^{\sigma_0+a+i\infty} - \int_{\sigma_0+a+i\epsilon}^{\sigma_0+a+i\infty} \phi'(s)\log Z(s)ds$$

$$= -\phi(\sigma_0 + a + i\epsilon)\log Z(\sigma_0 + a + i\epsilon)$$

$$- \sum_{q} c(q) \int_{\sigma_0+a+i\epsilon}^{\sigma_0+a+i\infty} \phi'(s)q^{-s}ds$$

Again, we can let ϵ approach zero to obtain the equality

$$\int_{\sigma_0+a}^{\sigma_0+a+i\infty} \phi(s)Z'/Z(s)ds$$

(5)
$$= -\phi(\sigma_0 + a)\log Z(\sigma_0 + a) - \sum_{q} c(q)\mathrm{Cr}_{q,a}^{+}(\phi).$$

The second term in (5) appears in Theorem 2.2. The first term in (5) does not appear in Theorem 2.2 because this term cancels with a term that appears in the evaluation of the bottom integral, as we shall now see.

In the evaluation of the bottom integral, we see the importance of choosing $\epsilon > 0$ before integrating by parts. By the choice of ϵ,

Z has no zeros or poles on the line segment $[-a + i\epsilon, \sigma_0 + a + i\epsilon]$, so we have

$$\int_{-a+i\epsilon}^{\sigma_0+a+i\epsilon} \phi(s)Z'/Z(s)ds$$

(6)
$$= \phi(s)\log Z(s)\Big|_{-a+i\epsilon}^{\sigma_0+a+i\epsilon} - \int_{-a+i\epsilon}^{\sigma_0+a+i\epsilon} \phi'(s)\log Z(s)ds.$$

Now let $\epsilon \to 0$ to get the equality

$$\int_{-a}^{\sigma_0+a} \phi(s)Z'/Z(s)ds$$

$$= \phi(\sigma_0 + a)\log Z(\sigma_0 + a) - \phi(-a)\log Z(-a)$$

(7)
$$- \int_{-a}^{\sigma_0+a} \phi'(s)\log Z(s)ds.$$

To complete the proof of Theorem 2.2, simply combine equations (2) through (7). Note the cancellation of one term in (5) with a term in (7).

Remark 1. The value of $\log Z(-a)$ is obtained by the analytic continuation of the Euler sum of Z along the horizontal line segment $[\sigma_0 + a + i\epsilon, -a + i\epsilon]$, followed by the continuation along the vertical line segment $[-a, -a + i\epsilon]$, which is equivalent to the analytic continuation along the top of the horizontal line segment $[-a, \sigma_0 + a]$ for small ϵ. To be precise, one should write the integral in (7) as

$$\int_{-a}^{\sigma_0+a} \phi'(s)\log Z(s)ds$$

(8)
$$= \int_{-a}^{\sigma_0+a} \phi'(s)\log |Z(s)|ds - \int_{-a}^{\sigma_0+a} \phi'(s)\arg(Z(s))ds.$$

Remark 2. In the case that $Z(s)$ is real on the real axis, then $\arg(Z(s))$ is a step function on $[-a, \sigma_0 + a]$ and takes on values in $\mathbf{Z} \cdot \pi i$, except at the zeros and poles of Z, where the argument is undetermined. In this case the integral with $\arg Z(s)$ in (8) can be evaluated directly and trivially, as an elementary integral.

§4. An inductive theorem.

When comparing our work with that of Cramér in [Cr 19], the reader should note that we have overcome a point of substantial technical difficulty that Cramér encountered when proving Theorem 2.2 for the Riemann zeta function $\zeta_{\mathbf{Q}}(s)$. By choosing a suitably, we have avoided having to consider the convergence of the Euler sum of Z on the line $\mathrm{Re}(s) = \sigma_0'$. Cramér used the fact that $\zeta_{\mathbf{Q}}(s)$ does not vanish on the vertical line $\mathrm{Re}(s) = 1$ as well as specific knowledge about the distribution of prime numbers, namely

$$\sum_{p \leq x} \frac{1}{p} = O(\log \log x) \quad \text{as } x \to \infty$$

and the Landau theorem which states that the limit

$$\lim_{x \to \infty} \sum_{p \leq x} \frac{1}{p^{1+it}}$$

converges uniformly for t in compact subsets of $\mathbf{R} \setminus \{0\}$. By following Cramér's original proof exactly, we would have greatly increased the complexity of the axioms of meromorphy, Euler sum, and functional equation.

The simplest example is that of the original Cramér theorem, for the Riemann zeta function $\zeta_{\mathbf{Q}}$. The gamma function obviously satisfies the growth conditions vis a vis the test function

$$\phi(s) = e^{sz} \quad \mathrm{Im}(z) > 0.$$

In fact, for the zeta functions coming from modular forms, with gamma factors as fudge factors, the same remark applies.

When applying Theorem 2.2 to the test function $\phi_z(s) = e^{sz}$, we can then set $z = it$, and we determined the asymptotic behavior of $\sum e^{i\rho t}$ as t approaches zero in complete detail in [JoL 93c] under the assumption that Φ is of regularized product type. The result of these calculations is the following theorem.

Theorem 4.1. *Let $(Z, \tilde{Z}, \Phi)$ be in the fundamental class, and assume that Φ has reduced order (M, m). Then Z and $\tilde{Z}$ are* of regularized product type and of reduced order $(M, m+1)$ if $-\mathrm{Re}(p_0) \in \mathbf{Z}$, otherwise the reduced order is (M, m).

Examples. In §7 of [JoL 93c] we applied Theorem 4.1 to give many examples of zeta functions which are thus sknown to be of regularized product type. We add to this list any **Dirichlet polynomial**, which we define to be any holomorphic function of the form

$$P(s) = \sum_{n=1}^{N} a_n b_n^s$$

where $\{a_n\}$ is a finite sequence of complex numbers, $\{b_n\}$ is a finite sequence of positive real numbers, which we may assume, without loss of generality, to satisfy the inequalities

$$0 < b_1 < b_2 < \cdots < b_N.$$

Let

$$Q(s) = P(-s) = \sum_{n=1}^{N} a_n b_n^{-s},$$

and write

$$P(s) = a_N b_N^s \left[1 + \sum_{n=1}^{N-1} \frac{a_n}{a_N} \left(\frac{b_n}{b_N} \right)^s \right] = a_N b_N^s \cdot Z(s)$$

and

$$Q(s) = a_1 b_1^{-s} \left[1 + \sum_{n=2}^{N} \frac{a_n}{a_1} \left(\frac{b_n}{b_1} \right)^{-s} \right] = a_1 b_1^{-s} \cdot \tilde{Z}(s).$$

It is immediate that there exists some $\sigma_0' > 0$ such that for all s with $\mathrm{Re}(s) > \sigma_0'$ we have

$$\left| \sum_{n=1}^{N-1} \frac{a_n}{a_N} \left(\frac{b_n}{b_N} \right)^s \right| < 1 \quad \text{and} \quad \left| \sum_{n=2}^{N} \frac{a_n}{a_1} \left(\frac{b_n}{b_1} \right)^{-s} \right| < 1.$$

Therefore, Z and $\tilde{Z}$ have Euler sums, which means there exist sequences $\{\mathbf{q}\}$, $\{\tilde{\mathbf{q}}\}$ and $\{c(\mathbf{q})\}$, $\{c(\tilde{\mathbf{q}})\}$ such that for $\mathrm{Re}(s) > \sigma_0'$ we have

$$\log Z(s) = \sum_{\mathbf{q}} \frac{c(\mathbf{q})}{\mathbf{q}^s} \quad \text{and} \quad \log \tilde{Z}(s) = \sum_{\tilde{\mathbf{q}}} \frac{c(\tilde{\mathbf{q}})}{\tilde{\mathbf{q}}^s}.$$

If we set
$$G(s) = a_N b_N^s \quad \text{and} \quad \tilde{G}(s) = a_1 b_1^{-s},$$

then the trivial relation $P(s) = Q(-s)$ can be written as

$$G(s)Z(s) = \tilde{G}(-s)\tilde{Z}(-s),$$

so we also have a functional equation with $\sigma_0 = 0$. Notice that the functional equation implies that all the zeros of P lie in some vertical strip. Further, we can apply Theorem 4.1 to conclude that the Dirichlet polynomial P is of regularized product type, with reduced order $(0, 1)$. As a result, the estimates from Chapter I, specifically Theorem 6.2, hold for any Dirichlet polynomial.

Finally, observe that the local factors of the more classical zeta functions are Dirichlet polynomials. Indeed, such factors are of the form $\text{Pol}_p(p^{-s})^{\pm 1}$ where Pol_p is a polynomial with constant term 1, and p is a prime number. For the Riemann zeta function, this local factor is simply $\text{Pol}_p(T) = 1 - T$, so we have $\text{Pol}_p(p^{-s}) = 1 - p^{-s}$. In the representation theory of $GL(n)$, the polynomial Pol_p has degree n. For representations in $GL(2)$ associated to an elliptic curve, say, we have $\text{Pol}_p(T) = 1 - a_p T + pT^2$, so in terms of p^{-s}, the local factor is

$$1 - a_p p^{-s} + pp^{-2s}.$$

Thus the local factors of classical zeta functions are themselves of regularized product type.

CHAPTER III

Explicit formulas under Fourier Assumptions

The classical "explicit formulas" of analytic number theory show that the sum of a certain function taken over the prime powers is equal to the sum of the Mellin transform taken over the zeros of the zeta function. Historically, only very special functions were used until Weil pointed out that the formulas could be proved for a much wider class of test functions (see [We 52]). We shall give here a version of these explicit formulas applicable to a wide class of test functions in connection with general zeta functions which have an Euler sum and functional equation whose fudge factor is of regularized product type. As a result, our general theorem contains the known explicit formulas for zeta functions of number fields and Selberg type zeta functions as well as new examples of explicit formulas such as that corresponding to the scattering determinant and Eisenstein series associated to any non-compact finite volume hyperbolic Riemann surface.

Various facts from analysis which we shall use in this chapter have been proved in our papers [JoL 93a] and [JoL 93b], as well as Chapter I. As a result, most of the steps taken here are relatively formal. We carry out the steps by integrating over a rectangle in the classical manner, but one aspect of this classical procedure emerges more clearly than in the case of classical zeta functions, namely the inductive procedure arising from a functional equation of the type

$$Z(s)\Phi(s) = \widetilde{Z}(\sigma_0 - s),$$

with zeta functions Z and $\widetilde{Z}$ and fudge factor Φ which is of regularized product type. For instance, for the Selberg zeta function of compact Riemann surfaces, these factors involve the Barnes double gamma function, and for the non-compact case, these fudge factors may involve the Riemann zeta function itself at the very least. Ultimately, arbitrarily complicated regularized products will occur as fudge factors in such a functional equation.

§1. Growth conditions on Fourier transforms.

We shall consider growth conditions on Fourier transforms and logarithmic derivatives of regularized products, and we begin by estimating Fourier transforms.

Following Barner [Ba 81], we require the test functions g to satisfy the following two basic Fourier conditions.

FOU 1. $g \in \mathrm{BV}(\mathbf{R}) \cap L^1(\mathbf{R})$.

FOU 2. g is **normalized**, meaning

$$g(x) = \frac{1}{2}\left(g(x+) + g(x-)\right) \quad \text{for all } x \in \mathbf{R}.$$

These will be the only relevant conditions in this section, but in the next section to apply a Parseval formula, we shall consider a third condition at the origin, namely:

FOU 3. There exists $\epsilon > 0$ such that

$$g(x) = g(0) + O(|x|^\epsilon) \quad \text{for } x \to 0.$$

If we let N be any integer ≥ 0, then we say that g satisfies the **basic Fourier conditions to order** N if g is N times differentiable and its first N derivatives satisfy the above three basic conditions.

Lemma 1.1. *Assume g satisfies* **FOU 1** *and* **FOU 2** *to order M. Then*
$$g^\wedge(t) = O(1/|t|^{M+1}) \quad \text{for } |t| \to \infty.$$

Proof. We integrate by parts M times to give

$$g^\wedge(t) = \frac{1}{\sqrt{2\pi}} \left(\frac{1}{it}\right)^M \int\limits_{-\infty}^{\infty} g^{(M)}(x)e^{-itx}\,dx.$$

To finish, note that for any $h \in \mathrm{BV}(\mathbf{R}) \cap L^1(\mathbf{R})$, we have the Stieltjes integration by parts formula

$$h^\wedge(t) = \frac{1}{\sqrt{2\pi}}\frac{1}{it} \int\limits_{-\infty}^{\infty} e^{-itx}\,dh(x),$$

from which we obtain the estimate

$$|h^\wedge(t)| \le \frac{1}{\sqrt{2\pi}|t|} V_{\mathbf{R}}(h),$$

whence the proposition follows. □

Let f be a measurable function on $\mathbf{R}^+$ so that under certain convergence conditions we have the **Mellin transform**

$$\mathbf{M}f(s) = \int\limits_0^\infty f(u)u^s \frac{du}{u}.$$

Let $\sigma_0 \in \mathbf{R}$. We define $\mathbf{M}_{\sigma_0/2}f$ to be the translate of $\mathbf{M}f$ by $\sigma_0/2$, meaning

$$\mathbf{M}_{\sigma_0/2}f(s) = \mathbf{M}f(s - \sigma_0/2).$$

We put

$$F(x) = f(e^{-x}) \quad \text{so} \quad f(u) = F(-\log u).$$

Then letting $s = \sigma + it$, we find

$$\mathbf{M}_{\sigma_0/2}f(s) = \int\limits_{-\infty}^\infty F(x)e^{x\sigma_0/2}e^{-sx}\,dx$$

$$= \int\limits_{-\infty}^\infty F(x)e^{-(\sigma-\sigma_0/2)x}e^{-itx}\,dx,$$

which is $\sqrt{2\pi}$ times the Fourier transform of

$$F_\sigma(x) = F(x)e^{-(\sigma-\sigma_0/2)x}.$$

That is,

$$\mathbf{M}_{\sigma_0/2}f(\sigma + it) = \sqrt{2\pi}F_\sigma^\wedge(t).$$

In particular, if we let $\sigma = \sigma_0/2$, then we obtain:

On the line $\mathrm{Re}(s) = \sigma_0/2$, *the Mellin transform is a constant multiple of the Fourier transform, namely*

$$\mathbf{M}_{\sigma_0/2}f(\sigma_0/2 + it) = \sqrt{2\pi}F^\wedge(t).$$

Lemma 1.2. *Let F be a function on $\mathbf{R}$ and assume there is an $\epsilon > 0$ such that*

$$F(x)e^{(\sigma_0/2-\sigma_1+\epsilon)|x|}.$$

Assume F satisfies **FOU 1** *and* **FOU 2** *to order M and define the function f on $\mathbf{R}^+$ by*

$$f(u) = F(-\log u).$$

Let $\sigma_1, \sigma_2 \in \mathbf{R}$ be fixed real numbers with $\sigma_1 < \sigma_2$ and consider the strip consisting of all $s \in \mathbf{C}$ with $\mathrm{Re}(s) \in [\sigma_1, \sigma_2]$. Then for any s in the strip $\mathrm{Re}(s) \in [\sigma_1, \sigma_2]$, we have

$$\mathbf{M}_{\sigma_0/2}f(\sigma + it) = \sqrt{2\pi}F_\sigma^\wedge(t).$$

and

$$\mathbf{M}_{\sigma_0/2}f(s) = O(1/|s|^{M+1})$$

for $|s| \to \infty$.

The proof of Lemma 1.2 follows that of Lemma 1.1, hence will be omitted.

As in previous chapters, we consider a meromorphic function H, and we want to compare the rate of growth of the logarithmic derivative H'/H and $M_{\sigma_0/2}f$ on vertical lines. Specifically, we consider the following growth condition.

GR. There exists a sequence $\{T_m\}$ tending to ∞ such that

$$M_{\sigma_0/2}f(s)H'/H(s) \to 0 \quad \text{for } m \to \infty$$

for any s on the horizontal line segment $S_{\pm m}$ defined by $\sigma \pm iT_m$ with $\sigma_1 \le \sigma \le \sigma_2$.

Remark. Of course, we are interested in considering the growth condition **GR** when H is one of the functions Z, $\widetilde{Z}$, or Φ. In the present variation, we need only one growth condition of type **GR** to insure the existence of certain limiting integrals. The point is that under the additional Fourier theoretic conditions, the convergence of the integrals on the vertical lines will be reduced to Fourier inversion on the middle, or critical, line $\mathrm{Re}(s) = \sigma_0/2$, after shifting the line of integration, and picking up residues corresponding to zeros and poles of the factor Φ.

In this chapter, we shall apply the estimates summarized in §6 of Chapter I to obtain a criterion under which the growth condition **GR** is satisfied, as in the next theorem.

Theorem 1.3. *Let H be of regularized product type of reduced order (M, m), and assume that F satisfies the growth conditions of Lemma 1.2 to order M. Then H satisfies the growth condition* **GR** *whenever F satisfies the three basic conditions to order M*

The proof of Theorem 1.3 follows directly from Theorem 2.5 of Chapter I and Lemma 1.2 above.

Recall that the basic Fourier conditions **FOU 1** and **FOU 2** insure the possibility of applying one of the most classical inversion theorems of Fourier analysis, stemming from Dirichlet, and attributed more directly to Pringsheim, Prasad and Hobson by Titchmarsch [Ti 48], page 25 (see also Theorem 2.5 of Chapter X in [La 93b]). For completeness, let us recall this result.

Basic Fourier Inversion Formula. *Let $\beta \in \mathrm{BV}(\mathbf{R}) \cap L^1(\mathbf{R})$, and suppose β is normalized. Then*

$$\lim_{T \to \infty} \frac{1}{\sqrt{2\pi}} \int_{-T}^{T} \beta^\wedge(t) e^{itx} \, dt = \beta(x) \quad \text{for all } x \in \mathbf{R}.$$

§2. The explicit formulas.

Let $(Z, \tilde{Z}, \Phi)$ be in the fundamental class. Let $a > 0$ be such that $\sigma_0 + a > \sigma_0'$ and such that Z, $\tilde{Z}$ and Φ do not have a zero or pole on the lines $\text{Re}(s) = -a$ and $\text{Re}(s) = \sigma_0 + a$. We let f and F be measurable functions related by

$$F(x) = f(e^{-x}) \quad \text{so} \quad f(u) = F(-\log u).$$

For the moment, assume there is some $a' > a$ for which we have the bound

(1) $$|F(x)| \ll e^{-(\sigma_0/2 + a')|x|}.$$

We shall actually assume something stronger later, but for now we just want to deal with a region of absolute convergence for a Mellin integral. Assuming the bound stated in (1), the function $M_{\sigma_0/2} f(s)$ is holomorphic in the closed strip $-a \leq \sigma \leq \sigma_0 + a$.

We let:

$\mathcal{R}_a$ be the infinite rectangle bounded by the vertical lines

$$\text{Re}(s) = -a \quad \text{and} \quad \text{Re}(s) = \sigma_0 + a.$$

$\mathcal{R}_a(T)$ be the finite rectangle bounded by the above vertical lines and the lines

$$\text{Im}(s) = -T \quad \text{and} \quad \text{Im}(s) = T.$$

We assume at first that Φ has no zeros or poles on the line $\text{Re}(s) = \sigma_0/2$. A variation without this restriction will also be treated. The line with $\sigma_0/2$ is especially useful for the more classical applications to counting **q**'s, or primes in the number theoretic case.

We let:

$\{\rho\}$ = the set of zeros and poles of Z in the *full* strip $-a \leq \sigma \leq \sigma_0 + a$;
$\{\alpha\}$ = the set of zeros and poles of Φ in the *half* strip $-a \leq \sigma \leq \sigma_0/2$;

Assume that T is chosen so that the functions Z, $\widetilde{Z}$, Φ have no zeros or poles on the horizontal lines that border $\mathcal{R}_a(T)$. Then, we may form the finite sum

$$V_{Z,a}(f,T) = \sum_{\rho \in \mathcal{R}_a(T)} v(\rho)\mathbf{M}_{\sigma_0/2}f(\rho)$$
$$= \mathrm{div}_{Z,a}(\mathbf{M}_{\sigma_0/2}f, T).$$

Similarly, we define

$$V_{\Phi,a,\sigma_0/2}(f,T) = \sum_{\alpha} v(\alpha)\mathbf{M}_{\sigma_0/2}f(\alpha).$$

Let $\mathcal{L}_a(T)$ be the boundary of the rectangle $\mathcal{R}_a(T)$ and consider the integral

$$\int_{\mathcal{L}_a(T)} \mathbf{M}_{\sigma_0/2}f(s)Z'/Z(s)ds.$$

By the residue theorem we have the equality

$$\int_{\mathcal{L}_a(T)} \mathbf{M}_{\sigma_0/2}f(s)Z'/Z(s)ds$$
$$= 2\pi i \sum_{\rho \in \mathcal{R}_a(T)} v(\rho)\mathbf{M}_{\sigma_0/2}f(\rho) = 2\pi i V_{Z,a}(f,T).$$

At this point, we want to take a limit as $T \to \infty$. For this, one must have a choice of $T \to \infty$ such that the integrand on the top and bottom segments of the rectangle tends to 0, as hypothesized in the growth condition **GR**. Roughly speaking, the more zeros and poles the function Z has in the strip $\mathcal{R}_a$, the larger $Z'/Z(s)$ could be on such horizontal segments, and so the smoother the function f must be so that its Mellin transform approaches 0 sufficiently fast, meaning faster than $Z'/Z(s)$ approaches infinity.

Assuming the growth condition **GR**, we are interested in the infinite sum

$$V_{Z,a}(f) = \sum_{\rho \in \mathcal{R}_a} v(\rho)\mathbf{M}_{\sigma_0/2}f(\rho),$$

which is understood in the limiting sense

$$(2) \quad \sum_{\rho \in \mathcal{R}_a} v(\rho) \mathbf{M}_{\sigma_0/2} f(\rho) = \lim_{m \to \infty} \sum_{\rho \in \mathcal{R}_a(T_m)} v(\rho) \mathbf{M}_{\sigma_0/2} f(\rho).$$

Since the similar sum over the family $\{\alpha\}$ is taken on the left half interval, we use the notation

$$V_{\Phi, a, \sigma_0/2}(f) = \sum_{\mathrm{Re}(\alpha) \leq \sigma_0/2} v(\alpha) \mathbf{M}_{\sigma_0/2} f(\alpha)$$

$$= \lim_{m \to \infty} \sum_{\alpha \in \mathcal{R}_a(T_m), \mathrm{Re}(\alpha) \leq \sigma_0/2} v(\alpha) \mathbf{M}_{\sigma_0/2} f(\alpha).$$

The Fourier conditions **FOU 1** and **FOU 2** are imposed in order to deal with questions of Fourier inversion in connection with sums over $\{q\}$. As we shall see, the condition **FOU 3** is concerned with our evaluation of the **Weil functional** W_Φ, which is defined to be the limit

$$W_\Phi(F) = \lim_{m \to \infty} \frac{1}{\sqrt{2\pi}} \int_{-T_m}^{T_m} F^\wedge(t) \Phi'/\Phi(\sigma_0/2 + it) dt,$$

where Φ is assumed to be holomorphic on the line $\mathrm{Re}(s) = \sigma_0/2$. In addition to these assumptions, we will require:

FOU 4. There exists a constant $a' > 0$ such that the function

$$x \mapsto F(x) e^{(\sigma_0/2 + a')|x|}$$

is in BV($\mathbf{R}$).

Under suitable conditions on the test function F, which will be expressed in terms of the above four Fourier conditions, we shall prove:

The Explicit Formula.

$$V_{Z,a}(f) + V_{\Phi, a, \sigma_0/2}(f) =$$

$$\sum_q \frac{-c(q) \log q}{q^{\sigma_0/2}} f(q) + \sum_{\tilde{q}} \frac{-c(\tilde{q}) \log \tilde{q}}{\tilde{q}^{\sigma_0/2}} f(1/\tilde{q}) + W_\Phi(F).$$

More specifically, the main result of this chapter is the following theorem.

Theorem 2.1. *Let $(Z, \tilde{Z}, \Phi)$ be in the fundamental class, and assume that Φ has reduced order (M, m) with no zeros or poles on the line $\mathrm{Re}(s) = \sigma_0/2$. Then for any function F which satisfies the four Fourier conditions to order M, the functionals $W_\Phi(F)$, $V_{Z,a}(f)$ and $V_{\Phi,a,\sigma_0/2}(f)$ are defined and the explicit formula holds.*

Observe that, as in our formulation of Cramér's theorem, the above theorem is an inductive one, expressing the sum $V_{Z,a}(f)$ in terms of a similar sum concerning $V_{\Phi,a,\sigma_0/2}$, the Weil functional, and terms involving the families $\{\mathbf{q}\}$ and $\{\tilde{\mathbf{q}}\}$.

Note that through a "change of notation", Theorem 2.1 can be used to express an explicit formula involving the zeros and poles of $\tilde{Z}$.

Remark 1. Suppose there are meromorphic functions G and $\tilde{G}$ such that

$$\Phi(s) = G(s)/\tilde{G}(\sigma_0 - s).$$

Then we may write the Weil functional as

$$W_\Phi(F) = W_G^+(F) + W_{\tilde{G}}^-(F)$$

where

$$W_G^+(F) = \lim_{m \to \infty} \frac{1}{\sqrt{2\pi}} \int_{-T_m}^{T_m} F^\wedge(t) G'/G(\sigma_0/2 + it) dt$$

and

$$W_{\tilde{G}}^-(F) = \lim_{m \to \infty} \frac{1}{\sqrt{2\pi}} \int_{-T_m}^{T_m} F^\wedge(-t) \tilde{G}'/\tilde{G}(\sigma_0/2 + it) dt.$$

Further, in the case where $G = \tilde{G}$, we can write the Weil functional as

$$W_\Phi(F) = \lim_{m \to \infty} \frac{1}{\sqrt{2\pi}} \int_{-T_m}^{T_m} [F^\wedge(t) + F^\wedge(-t)] G'/G(\sigma_0/2 + it) dt.$$

Remark 2. Observe that the sum $V_{Z,a} + V_{\Phi,a,\sigma_0/2}$ is independent of a even though neither term is independent of a.

For any u such that Φ has no zero or pole on the line $\mathrm{Re}(s) = u$, we define

$$W_{\Phi,u}^{\#}(f) = \lim_{m \to \infty} \frac{1}{2\pi i} \int_{u-iT_m}^{u+iT_m} \mathbf{M}_{\sigma_0/2}(f)\Phi'/\Phi(s)ds.$$

The proof of the explicit formula will go through the following intermediate stage.

Theorem 2.2. *Assume that Φ is of regularized product type of reduced order (M, m), and F satisfies the Fourier conditions to order M. Then*

$$V_{Z,a}(f) = \sum_{\mathbf{q}} \frac{-c(\mathbf{q})\log \mathbf{q}}{\mathbf{q}^{\sigma_0/2}} f(\mathbf{q}) + \sum_{\tilde{\mathbf{q}}} \frac{-c(\tilde{\mathbf{q}})\log \tilde{\mathbf{q}}}{\tilde{\mathbf{q}}^{\sigma_0/2}} f(1/\tilde{\mathbf{q}})$$

$$+ W_{\Phi,-a}^{\#}(f).$$

Theorem 2.2 will be proved in §3. After Theorem 2.2 has been proved, what will remain is to analyze the last term containing $\Phi'/\Phi(s)$. Different applications require different analyses of this term. For classical analytic number theory, one moves the line of integration from $-a$ to $\sigma_0/2$, and then one applies the general Parseval formula. This will be carried out in §4, thus yielding Theorem 2.1. In §5 we give a further determination of the Weil functional, based on the general Parseval formula from [JoL 93b].

However, for other applications and notably those occuring later in this book in Chapter V, §3 and §4, we shall move the line of integration far to the right. In these applications, one can thus completely bypass the Parseval formula, and the final result is therefore much simpler to prove. We now carry this out.

Any regularized product type can be expressed as a product of two such types, one of which has all of its zeros and poles in a left half plane and one of which has all of its zeros and poles in a right half plane, say

$$\Phi = \Phi_{\text{left}} \Phi_{\text{right}}.$$

We let $A > \sigma_0$ be a number such that all the zeros and poles of Φ_{left} are in the half plane $\text{Re}(s) \leq A - \delta$, for some $\delta > 0$. Similarly, select a is such that all the zeros and poles of Φ_{right} are in the half plane $\text{Re}(s) \geq -a + \delta$, for some $\delta > 0$.

In Theorem 2.2 we move the line of integration for Φ_{left} to $\text{Re}(s) = A$, thus picking up the residues of all the poles between these two lines. At the same time, we define

$$V_{\Phi_{\text{left}},a,A}(f) = \sum_{\zeta} v(\zeta) \mathbf{M}_{\sigma_0/2} f(\zeta)$$

where

$\{\zeta\}$ = the set of all zeros and poles of Φ_{left} such that $\text{Re}(\zeta) > -a$.

Then, from Theorem 2.2, we arrive at the following formula.

Theorem 2.3. *Let $(Z, \tilde{Z}, \Phi)$ be in the fundamental class, and assume that Φ is decomposed as above. Then for any function F which satisfies the four Fourier conditions to order M, all the functionals in the next formula are defined and the following formula holds:*

$$V_{Z,a}(f) + V_{\Phi_{\text{left}},a,A}(f) =$$

$$\sum_{\mathbf{q}} \frac{-c(\mathbf{q}) \log \mathbf{q}}{\mathbf{q}^{\sigma_0/2}} f(\mathbf{q}) + \sum_{\tilde{\mathbf{q}}} \frac{-c(\tilde{\mathbf{q}}) \log \tilde{\mathbf{q}}}{\tilde{\mathbf{q}}^{\sigma_0/2}} f(1/\tilde{\mathbf{q}})$$

$$+ W^{\#}_{\Phi_{\text{left}},A}(f) + W^{\#}_{\Phi_{\text{right}},-a}(f).$$

Steps which justify moving the line of integration are given in §4 in the case considered in Theorem 2.1. The same argument applies to Theorem 2.2, thus yielding Theorem 2.3.

Remark 3. In Chapter II, for the proof of Cramèr's theorem, we did not use a Mellin transform but worked directly on a half strip with a test function ϕ. One may do the same in a full strip to get the explict formula directly for such a function. In that case, the Weil functionals are expressed in terms of ϕ instead of the Mellin transform $\mathbf{M}_{\sigma_0/2} f$. For the applications to Chapter IV and Chapter V, this way of proceeding eliminates completely all Fourier conditions, and we could deal directly with the special test functions

$$\phi_t(s) = e^{(s-\sigma_0/2)^2 t}$$

72

or

$$\phi_t'(s) = (2s - \sigma_0)e^{(s-\sigma_0/2)^2 t},$$

just as we dealt with the function e^{sz} for the Cramér theorem in [JoL 93c].

§3. The terms with the q's.

In this and the next section we will begin our proof of the explicit formula based on the axioms **FOU 1** through **FOU 4**. In this section, we shall compute the sums over **q** that appear in Theorem 2.1. The remainder of the proof, namely the determination of the terms involving the fudge factor Φ and its set of zeros and poles $\{\alpha\}$ will be given in the next section.

Consider the rectangle $\mathcal{R}_a(T_m)$, as defined above, and integrate over the boundary of this rectangle. By the residue formula, we have

$$(1) \qquad \sum_{\rho \in \mathcal{R}_a(T_m)} v(\rho) \mathbf{M}_{\sigma_0/2} f(\rho) = \frac{1}{2\pi i} \int_{\mathcal{L}_a(T_m)} \mathbf{M}_{\sigma_0/2} f(s) Z'/Z(s) ds.$$

Throughout we assume that Φ is of regularized product type of reduced order (M, m). By Theorem 4.1 of Chapter II, Z and $\widetilde{Z}$ are of regularized product type of reduced order $(M, m+1)$. We now can apply Lemma 1.2 to $H = Z$ to find, for $m \to \infty$,

$$\frac{1}{2\pi i} \int_{-a-iT_m}^{\sigma_0+a-iT_m} \mathbf{M}_{\sigma_0/2} f(s) Z'/Z(s) ds$$

$$(2) \qquad + \frac{1}{2\pi i} \int_{\sigma_0+a+iT_m}^{-a+iT_m} \mathbf{M}_{\sigma_0/2} f(s) Z'/Z(s) ds = o(1).$$

Upon combining (1) and (2), we obtain

$$\sum_{\rho \in \mathcal{R}_a(T_m)} v(\rho) \mathbf{M}_{\sigma_0/2} f(\rho) + o(1)$$

$$= \frac{1}{2\pi i} \int_{-a+iT_m}^{-a-iT_m} \mathbf{M}_{\sigma_0/2} f(s) Z'/Z(s) ds$$

$$+ \frac{1}{2\pi i} \int_{\sigma_0+a-iT_m}^{\sigma_0+a+iT_m} \mathbf{M}_{\sigma_0/2} f(s) Z'/Z(s) ds.$$

Using the functional equation, we can express this equality as

$$
= \frac{1}{2\pi i} \int_{-a+iT_m}^{-a-iT_m} \mathbf{M}_{\sigma_0/2} f(s) \left[-\tilde{Z}'/\tilde{Z}(\sigma_0 - s) \right] ds
$$

(3)
$$
+ \frac{1}{2\pi i} \int_{-a+iT_m}^{-a-iT_m} \mathbf{M}_{\sigma_0/2} f(s) \left[-\Phi'(s)/\Phi(s) \right] ds
$$

$$
+ \frac{1}{2\pi i} \int_{\sigma_0+a-iT_m}^{\sigma_0+a+iT_m} \mathbf{M}_{\sigma_0/2} f(s) Z'/Z(s) ds.
$$

As stated above, in this section we will deal with the Z and $\tilde{Z}$ integrals, and in the next section we will evaluate the Φ integral.

For the Z integral, we obtain

$$
\frac{1}{2\pi i} \int_{\sigma_0+a-iT_m}^{\sigma_0+a+iT_m} \mathbf{M}_{\sigma_0/2} f(s) Z'/Z(s) ds
$$

(4)
$$
= \frac{-1}{2\pi} \int_{-T_m}^{T_m} dt \int_{-\infty}^{\infty} F(x) e^{-(\sigma_0/2+a+it)x} \sum_{q} \frac{c(q) \log q}{q^{\sigma_0+a+it}} dx.
$$

This step follows by differentiating the Euler sum for $\log Z$ to obtain the series for Z'/Z. We shall give formal arguments to change this last expression (4), and then we give estimates to justify the calculations. We make a change of variables

$$
x = y - \log q \quad \text{and} \quad dx = dy
$$

in the integral of each term with q. Then the whole expression (4) becomes

$$
= \frac{-1}{2\pi} \int_{-T_m}^{T_m} dt \sum_{q} \int_{-\infty}^{\infty} \frac{c(q) \log q}{q^{\sigma_0/2}} F(y - \log q) e^{-(\sigma_0/2+a)y} e^{-ity} dy
$$

$$
= \frac{-1}{2\pi} \int_{-T_m}^{T_m} dt \sum_{q} \int_{-\infty}^{\infty} H_q(y) e^{-ity} dy
$$

where

$$H_{\mathbf{q}}(y) = \frac{c(\mathbf{q}) \log \mathbf{q}}{\mathbf{q}^{\sigma_0/2}} F(y - \log \mathbf{q}) e^{-(\sigma_0/2 + a)y}.$$

We let

$$H(y) = \sum_{\mathbf{q}} H_{\mathbf{q}}(y).$$

With the definition of H, we may express our desired limit integral in (4) as being

$$= \frac{-1}{2\pi} \lim_{m \to \infty} \int_{-T_m}^{T_m} dt \int_{-\infty}^{\infty} H(y) e^{-ity} \, dy = -H^{\wedge\wedge}(0) = -H(0).$$

by the Basic Fourier Inversion Formula. We then see that

$$-H(0) = \sum_{\mathbf{q}} \frac{-c(\mathbf{q}) \log \mathbf{q}}{\mathbf{q}^{\sigma_0/2}} F(-\log \mathbf{q}) = \sum_{\mathbf{q}} \frac{-c(\mathbf{q}) \log \mathbf{q}}{\mathbf{q}^{\sigma_0/2}} f(\mathbf{q}),$$

which is a term in the statement of Theorem 2.1.

We shall now justify the above steps. First, by **FOU 4**, there is a constant C such that

$$|F(x)| \le C e^{-(\sigma_0/2 + a')|x|},$$

from which we get the estimates

$$|F(y - \log \mathbf{q})| \, e^{-(\sigma_0/2 + a)y} \le \begin{cases} C \mathbf{q}^{-(\sigma_0/2 + a')} e^{(a' - a)y} & y \le \log \mathbf{q} \\ C \mathbf{q}^{\sigma_0/2 + a'} e^{-(\sigma_0 + a + a')y} & y \ge \log \mathbf{q}, \end{cases}$$

and, in particular,

$$|F(y - \log \mathbf{q})| \, e^{-(\sigma_0/2 + a)y} \le C \mathbf{q}^{-(\sigma_0/2 + a)}$$

for all y. From this it follows that

(5)
$$|H_{\mathbf{q}}(y)| \le 2C \frac{|c(\mathbf{q})| \log \mathbf{q}}{\mathbf{q}^{\sigma_0 + a}},$$

and after actually performing the integration, we obtain

$$(6) \qquad \int\limits_{-\infty}^{\infty} |H_{\mathbf{q}}(y)|\, dy \le 2C \frac{|c(\mathbf{q})|\log \mathbf{q}}{\mathbf{q}^{\sigma_0+a}} \left[\frac{1}{a'-a} + \frac{1}{\sigma_0+a+a'} \right].$$

Estimate (5) shows that the series

$$H(y) = \sum_{\mathbf{q}} H_{\mathbf{q}}(y)$$

is absolutely and uniformly convergent, and estimate (6) shows that this series defines a function $y \mapsto H(y)$ in $L^1(\mathbf{R})$.

Since each term $H_{\mathbf{q}}$ is of bounded variation and normalized, the uniform convergence of the series (6) shows that H is normalized. Furthermore, the total variation $V_{\mathbf{R}}$ satisfies the triangle inequality for an infinite sum, as one verifies directly from the Riemann-Steiltjes sums defining this variation. Similarly for a product, we have

$$V_{\mathbf{R}}(gh) \le \|g\| V_{\mathbf{R}}(h) + \|h\| V_{\mathbf{R}}(g).$$

Then

$$V_{\mathbf{R}}(H) \le \sum_{\mathbf{q}} \frac{|c(\mathbf{q})|\log \mathbf{q}}{\mathbf{q}^{\sigma_0+a}} V_{\mathbf{R}}\left(F(x)e^{-(\sigma_0/2+a)x} \right)$$

Hence, by **FOU 4**, H has bounded variation, and we have justified all the formal operations and the application of Fourier inversion in the evaluation of the Z integral as $-H(0)$.

One may carry out similar arguments for the $\widetilde{Z}$ integral

$$\frac{-1}{2\pi i} \int\limits_{-a+iT_m}^{-a-iT_m} \mathbf{M}_{\sigma_0/2} f(s)\widetilde{Z}'/\widetilde{Z}(\sigma_0 - s)\, ds,$$

which we write as

$$= \frac{-1}{2\pi} \int\limits_{-T_m}^{T_m} dt \int\limits_{-\infty}^{\infty} F(x)e^{-(\sigma_0/2+a+it)x} \sum_{\tilde{\mathbf{q}}} \frac{c(\tilde{\mathbf{q}})\log \tilde{\mathbf{q}}}{\tilde{\mathbf{q}}^{\sigma_0+a+it}}\, dx.$$

In this case, one uses the inequalities

$$|F(y + \log \tilde{\mathbf{q}})| \, e^{(\sigma_0/2+a)y} \leq \begin{cases} C\tilde{\mathbf{q}}^{-(\sigma_0/2+a')} e^{(a-a')y} & y \geq -\log \tilde{\mathbf{q}} \\ C\tilde{\mathbf{q}}^{\sigma_0/2+a'} e^{(\sigma_0+a+a')y} & y \leq -\log \tilde{\mathbf{q}}, \end{cases}$$

and

$$|F(y + \log \tilde{\mathbf{q}})| \, e^{(\sigma_0/2+a)y} \leq C\tilde{\mathbf{q}}^{-(\sigma_0/2+a)},$$

which holds for all y. Inequalities similiar to (5) and (6) then follow when we define

$$H_{\tilde{\mathbf{q}}}(y) = \frac{c(\tilde{\mathbf{q}}) \log \tilde{\mathbf{q}}}{\tilde{\mathbf{q}}^{\sigma_0/2}} F(y + \log \tilde{\mathbf{q}}) e^{(\sigma_0/2+a)y}$$

and

$$H(y) = \sum_{\tilde{\mathbf{q}}} H_{\tilde{\mathbf{q}}}(y),$$

with the bounded variation

$$V_{\mathbf{R}}(H) \leq \sum_{\tilde{\mathbf{q}}} \frac{|c(\tilde{\mathbf{q}})| \log \tilde{\mathbf{q}}}{\tilde{\mathbf{q}}^{\sigma_0+a}} V_{\mathbf{R}} \left(F(x) e^{(\sigma_0/2+a)x} \right).$$

In this way, we obtain the term involving the $\widetilde{Z}$ integral in Theorem 2.1. This concludes the proof of the Theorem 2.1 as far as it involves the Z and $\widetilde{Z}$ integrals.

At this point, we have proved Theorem 2.2.

§4. The term involving Φ.

In this section we will compute the terms in the explicit formula containing the fudge factor Φ and its set of roots $\{\alpha\}$. To do so, we begin by considering the integral containing Φ'/Φ, namely

$$\frac{1}{2\pi i} \int_{-a-iT_m}^{-a+iT_m} \mathbf{M}_{\sigma_0/2} f(s)\Phi'/\Phi(s)ds.$$

We want to move the line of integration to the critical line $\sigma = \sigma_0/2$. Upon doing this, we pick up the residues of $\mathbf{M}_{\sigma_0/2} f(s)\Phi'/\Phi(s)$ at the points α, and, hence, by using **FOU 1**, we find

$$\frac{1}{2\pi i} \int_{-a-iT_m}^{-a+iT_m} \mathbf{M}_{\sigma_0/2} f(s)\Phi'/\Phi(s)ds$$

$$= \frac{1}{2\pi i} \int_{\sigma_0/2-iT_m}^{\sigma_0/2+iT_m} \mathbf{M}_{\sigma_0/2} f(s)\Phi'/\Phi(s)ds$$

$$- \sum_{\alpha\in\mathcal{R}_a(T_m)} v(\alpha)\mathbf{M}_{\sigma_0/2} f(\alpha) + o(1)$$

$$= \frac{1}{\sqrt{2\pi}} \int_{-T_m}^{T_m} F^\wedge(t)\Phi'/\Phi(\sigma_0/2 + it)dt$$

(1)

$$- \sum_{\alpha\in\mathcal{R}_a(T_m)} v(\alpha)\mathbf{M}_{\sigma_0/2} f(\alpha) + o(1).$$

Using **FOU 1** and **FOU 2**, we obtain the equality

$$\frac{1}{2\pi i} \lim_{m\to\infty} \int_{-a-iT_m}^{-a+iT_m} \mathbf{M}_{\sigma_0/2} f(s)\Phi'/\Phi(s)ds = W_\Phi(F) - V_{\Phi,a,\sigma_0/2}(f).$$

With this, as well as the calculations from the previous section, we have concluded the proof of Theorem 2.1, up to the evaluation of the Weil functional, which will now be dealt with.

§5. The Weil functional and regularized product type.

In this section we consider the evaluation of the Weil functional W_Φ when Φ is of regularized product type of reduced order (M, m) and F satisfies the Fourier conditions to order M. Essentially, we will apply the general Parseval formula of [JoL 93b].

Quite generally, for suitable functions g and H, and $u \in \mathbf{R}$, we shall consider the Weil functional, which we define to be

$$W_{H,u}(g) = \lim_{T \to \infty} \frac{1}{\sqrt{2\pi}} \int_{-T}^{T} g^{\wedge}(t) H'/H(u + it)dt.$$

Our analysis involves cases when H is one of the following three types of functions.

Case 1. $H = Q$ for some rational function Q.

Case 2. $H = e^P$ for some polynomial P.

Case 3. $H = D$ for some regularized product D.

In the remainder of this section, we will evaluate the Weil functional in each of the above three cases.

Case 1: $H = Q$ for some rational function Q. It suffices to consider the function $H(z) = z + \alpha$ for some complex number α.

Theorem 5.1. *Assume g satisfies* **FOU 1** *and* **FOU 2**. *Then*

$$\frac{1}{\sqrt{2\pi}} \int_{-\infty}^{\infty} g^{\wedge}(t) \frac{1}{t + \alpha} dt = \begin{cases} -i \int_0^{\infty} g(x) e^{i\alpha x} dx, & \mathrm{Im}(\alpha) > 0, \\ i \int_0^{\infty} g(-x) e^{-i\alpha x} dx, & \mathrm{Im}(\alpha) < 0. \end{cases}$$

Proof. Lemma 1.1 shows that $g^{\wedge}(t)/(t + \alpha)$ is in $L^1(\mathbf{R})$, so if

$\text{Im}(\alpha) > 0$, we have

$$\frac{1}{\sqrt{2\pi}} \int\limits_{-\infty}^{\infty} g^\wedge(t)\frac{1}{t+\alpha}\,dt = \frac{1}{\sqrt{2\pi}} \int\limits_{-\infty}^{\infty} g^\wedge(t) \int\limits_{0}^{\infty} -ie^{i(t+\alpha)x}\,dx\,dt$$

$$= -i \int\limits_{0}^{\infty} g(x)e^{i\alpha x}\,dx,$$

by the Fubini Theorem and the Basic Fourier Inversion Formula. The case $\text{Im}(\alpha) < 0$ is treated similarly. $\quad\square$

Case 2: $H = e^P$ for some polynomial P.

In this case, the evaluation of the Weil functional reduces to the Basic Fourier Inversion Formula as the following theorem demonstrates.

Theorem 5.2. *Assume g satisfies* **FOU 1** *and* **FOU 2** *to order M. Let P' be a polynomial of degree $\leq M$, and let $P'(-i\partial)$ be the corresponding constant coefficient partial differential operator. Then*

$$\lim_{T \to \infty} \frac{1}{\sqrt{2\pi}} \int\limits_{-T}^{T} g^\wedge(t)P'(t)\,dt = P'(-i\partial)g(0).$$

Proof. The Basic Fourier Inversion Formula states

$$\lim_{T \to \infty} \frac{1}{\sqrt{2\pi}} \int\limits_{-T}^{T} g^\wedge(t)P'(t)e^{itx}\,dt = P'(-i\partial)g(x).$$

The assertion follows after we set $x = 0$. $\quad\square$

Case 3: $H = D$ for some regularized product D.

This case is handled by the results from [JoL 92b], which generalizes Barner's formulation [Ba 81] of Weil's formula [We 52] in the special case H'/H is the logarithmic derivative of the classical gamma function. For completeness, let us recall the main theorem from [JoL 92b].

Let L and A be as in §1, and let $\theta(x)$ be the corresponding theta function

$$\theta(x) = \sum a_k e^{-\lambda_k t}$$

that satisfies the asymptotic axioms **AS 1**, **AS 2** and **AS 3**. For any n let

$$L_n = \{\lambda_{n+1}, \dots, \},$$

so then we have

(1)
$$D'_L/D_L(z) = \sum_{k=0}^{n} \frac{a_k}{z + \lambda_k} + D'_{L_n}/D_{L_n}(z).$$

Since Theorem 5.1 applies to the sum in (1), we may assume, without loss of generality, that L is such that

$$\max_{\lambda_k \in L}\{-\operatorname{Re}(\lambda_k)\} < u.$$

The principal part of the theta function $\theta(x)$ is

$$P_0\theta(x) = \sum_{\operatorname{Re}(p)<0} b_p(x)x^p,$$

which, by **AS 2**, is such that one has the asymptotic behavior

$$\theta(x) - P_0\theta(x) = O(|\log x|^{m(0)}) \quad \text{as } x \to 0.$$

For any $z \in \mathbf{C}$, let

$$\theta_z(x) = \theta(x)e^{-zx}.$$

By expanding e^{-zx} in a power series, we see that the principal part of $\theta_z(x)$ is

$$P_0\theta_z(x) = P_0\left[e^{-zx}\theta(x)\right] = \sum_{\operatorname{Re}(p)+k<0} \frac{b_p(x)x^{p+k}}{k!}(-z)^k.$$

Recall from §1 (Theorem 4.1 of [JoL 92a]) that, for any constant $\alpha > 0$, the logarithmic derivative of the regularized product can be written as

$$D'/D(u + it) = I_\alpha(u - \alpha + it) + S_\alpha(u - \alpha + it)$$

where

$$I_w(z) = \int_0^\infty [\theta_z(x) - P_0\theta_z(x)]\, e^{-wx}\, dx$$

and

$$S_w(z) = \sum_{\mathrm{Re}(p)+k<0} \frac{(-z)^k}{k!} CT_{s=0} B_p(\partial_s) \left[\frac{\Gamma(s + p + k + 1)}{w^{s+p+k+1}} \right].$$

Note that, as a function of z, $S_\alpha(z)$ is a polynomial of degree $\leq M$, in which case Theorem 5.2 applies to show that $S_\alpha(z)$ satisfies **GR** for any test function g that satisfies the basic conditions to order M. Therefore, in order to evaluate

$$\lim_{T\to\infty} \frac{1}{\sqrt{2\pi}} \int_{-T}^{T} g^\wedge(t) D'/D(u + it)\, dt,$$

it suffices to evaluate

$$\lim_{T\to\infty} \frac{1}{\sqrt{2\pi}} \int_{-T}^{T} g^\wedge(t) I_\alpha(u - \alpha + it)\, dt.$$

If we restrict the complex variable z to a vertical line by setting $z = u - \alpha + it$, we get

$$P_0\theta_z(x) = \sum_{\mathrm{Re}(p)+k<0} \frac{b_p(x)x^{p+k}}{k!}(-u + \alpha - it)^k$$

$$= \sum_{k<-\mathrm{Re}(p_0)} c_k(u - \alpha, x)(-it)^k,$$

where the coefficients $c_k(u - \alpha, x)$ depend on the variables $u - \alpha$ and x through the coefficients of t^p for $\mathrm{Re}(p) < 0$. With this, we have

$$I_\alpha(u - \alpha + it)$$

$$= \int\limits_0^\infty \left[\theta(x)e^{-x(u-\alpha+it)} - \sum_{k < -\mathrm{Re}(p_0)} c_k(u - \alpha, x)(-it)^k \right] e^{-\alpha x} dx.$$

Let us define

$$d\mu_\alpha(x) = e^{-\alpha x} dx \quad \text{and} \quad \theta_{u-\alpha}(x) = \theta(x)e^{-(u-\alpha)x}.$$

Therefore, the above calculations yield the equality

$$I_\alpha(u - \alpha + it)$$

$$= \int\limits_0^\infty \left[\theta_{u-\alpha}(x)e^{-itx} - \sum_{k < -\mathrm{Re}(p_0)} c_k(u - \alpha, x)(-it)^k \right] d\mu_\alpha(x),$$

Applying Theorem 4.3 from [JoL 92b] we obtain:

Theorem 5.3. *Assume g satisfies the Fourier conditions to order M. Then, with notation as above, we have*

$$\lim_{T \to \infty} \frac{1}{\sqrt{2\pi}} \int\limits_{-T}^T g^\wedge(t) I_\alpha(u - \alpha + it) dt$$

$$= \int\limits_0^\infty \left[\theta_{u-\alpha}(x)g(-x) - \sum_{k < -\mathrm{Re}(p_0)} c_k(u - \alpha, x)g^{(k)}(0) \right] d\mu_\alpha(x).$$

In summary, we have shown:

Theorem 5.4. *Assume that Φ is of regularized product type or reduced order M. If g is a test function satisfying the four basic Fourier conditions to order M, then the Weil functional $W_{\Phi,u}(g)$ is defined. Further, Theorems 5.1, 5.2 and 5.3 combine to provide an explicit evaluation of this functional.*

For the sake of space, we will explicitly evaluate the Weil functional only for special functions Φ, namely those involving the classical gamma function.

Example: The gamma function. Many known zeta functions, such as those associated to number fields or modular forms, have Euler sums and functional equations with fudge factors which involve the gamma function. For example, the zeta function corresponding to an ideal class in a number field k has a functional equation with $\sigma_0 = 1$ and

$$G(s) = \tilde{G}(s) = A^{s/2}\Gamma(s/2)^{r_1}\Gamma(s)^{r_2},$$

so

$$G'/G(s) = \frac{1}{2}\log A + \frac{r_1}{2}\Gamma'/\Gamma(s/2) + r_2\Gamma'/\Gamma(s).$$

So, the evaluation of the Weil functional in this case reduces to considering the logarithmic derivative of the gamma function. As recalled on page 52 of [JoL 93a] and §1 of Chapter I, the (lesser known) Gauss formula for the gamma function states that for any $z \in \mathbf{C}$ with $\mathrm{Re}(z) > -1$, we have

$$\Gamma'/\Gamma(z+1) = -\int_0^\infty \left[e^{-zt}\theta(t) - \frac{1}{t}\right]e^{-t}dt$$

where

$$\theta(t) = \sum_{n=0}^\infty e^{-nt} = \frac{1}{1 - e^{-t}}.$$

From this, and Theorem 4.3 of [JoL 93b], for any f which satisfies the four basic Fourier conditions and $a > -1$, we have

$$-\lim_{T\to\infty}\frac{1}{\sqrt{2\pi}}\int_{-T}^{T} f^\wedge(t)\Gamma'/\Gamma(a + it + 1)dt$$

$$= \int_0^\infty \left[f(-x)e^{-ax}\theta(x) - \frac{f(0)}{x}\right]e^{-x}dx.$$

This is Weil's formula as reworked by Barner (see [Ba 81], page 146).

CHAPTER IV

From Functional Equations to Theta Inversions

The classical Jacobi theta inversion formula for the Riemann theta function of one variable states that for all $t > 0$, we have the equality

$$\frac{1}{2\pi} \sum_{n=-\infty}^{\infty} e^{-n^2 t} = \frac{1}{\sqrt{4\pi t}} \sum_{n=-\infty}^{\infty} e^{-(2\pi n)^2/4t}.$$

If we set

$$\theta(u) = \sum_{n=-\infty}^{\infty} e^{-n^2 \pi u},$$

then the Jacobi inversion formula can be stated as the equality

$$\theta(u) = \frac{1}{\sqrt{u}} \theta(1/u) \quad \text{for } u > 0.$$

Spectrally, the inversion formula can be viewed as expressing a sum over all the eigenvalues of the Laplace operator on the circle (namely the squares of the integers) as equal to another similar sum, with the inversion $t \mapsto 1/t$.

We give the following very simple spectral interpretation of the Jacobi inversion formula. Let $X = 2\pi \mathbf{Z} \backslash \mathbf{R}$ be the circle. The heat kernel for the usual Laplacian on $\mathbf{R}$ is

$$K_{\mathbf{R}}(x, t, y) = \frac{1}{\sqrt{4\pi t}} e^{-(x-y)^2/4t}.$$

The heat kernel on X is the $2\pi \mathbf{Z}$ periodization of the heat kernel $K_{\mathbf{R}}$ on $\mathbf{R}$. On the other hand, the eigenfunction expansion of the heat kernel K_X can be easily computed. When we equate this periodization with the eigenfunction expansion of the heat kernel, we obtain what amounts to a theta inversion formula, namely

$$\frac{1}{\sqrt{4\pi t}} \sum_{n=-\infty}^{\infty} e^{-(x-y+2\pi n)^2/4t} = \frac{1}{2\pi} \sum_{n=-\infty}^{\infty} e^{-n^2 t} e^{inx} e^{-iny}.$$

In Theorem 1.1 below, we show how the above classical theta inversion formula admits a vast extension to much more general theta functions, essentially formed with the sequence of zeros and poles of functions in the fundamental class as defined in Chapter II, §1. Specifically, inversion formulas follow from our general explicit formula when using Gaussian type test functions. In this context, the Jacobi inversion formula comes from the explicit formulas associated to the sine function

$$\sin(\pi i s) = \frac{e^{-\pi s} - e^{\pi s}}{2i} = -\frac{e^{\pi s}}{2i}\left(1 - e^{-2\pi s}\right)$$

which, when written in this form, can be seen to have an Euler sum and functional equation with $\sigma_0 = 0$ and a simple exponential fudge factor. We will prove our general inversion formulas in §1, and give various examples in §2.

We will show conversely in Chapter V how inversion formulas for theta functions satisfying **AS 1**, **AS 2**, and **AS 3** yield Dirichlet series with an additive functional equation.

§1. An application of the explicit formulas.

We shall apply the general explicit formula of Chapter III, to the test function f_t defined for $t > 0$ by

$$(1) \quad f_t(u) = \frac{1}{\sqrt{4\pi t}} e^{-(\log u)^2/4t} \quad \text{so} \quad F_t(x) = \frac{1}{\sqrt{4\pi t}} e^{-x^2/4t}.$$

Note that F_t is the heat kernel on $\mathbf{R}$. It is immediate that F_t satisfies the four basic Fourier conditions needed in the proof of the explicit formulas. By a direct calculation, we have

$$(2) \quad \mathbf{M}_{\sigma_0/2} f_t(s) = e^{(s-\sigma_0/2)^2 t} \quad \text{and} \quad F_t^\wedge(r) = \frac{1}{\sqrt{2\pi}} e^{-r^2 t}.$$

For example, to derive the first formula in (2), write

$$\mathbf{M}_{\sigma_0/2} f_t(s) = \int\limits_0^\infty f_t(u) u^{s-\sigma_0/2} \frac{du}{u}$$

$$= \frac{1}{\sqrt{4\pi t}} \int\limits_{-\infty}^\infty e^{-x^2/4t + x(s-\sigma_0/2)} dx$$

$$= e^{(s-\sigma_0/2)^2 t}.$$

With this, the explicit formula yields the following result, which we call a **theta inversion formula.**

Theorem 1.1. *Let $(Z, \tilde{Z}, \Phi)$ be in the fundamental class, and assume that Φ has no zeros or poles on the line $\mathrm{Re}(s) = \sigma_0/2$. Let $\{\rho\}$ and $\{\alpha\}$ be as in Chapter III, §2. Then for all $t > 0$ we have*

$$\sum_\rho v(\rho) e^{(\rho-\sigma_0/2)^2 t} + \sum_\alpha v(\alpha) e^{(\alpha-\sigma_0/2)^2 t}$$

$$= \frac{1}{\sqrt{4\pi t}} \left[\sum_{\mathbf{q}} \frac{-c(\mathbf{q}) \log \mathbf{q}}{\mathbf{q}^{\sigma_0/2}} e^{-(\log \mathbf{q})^2/4t} \right]$$

$$+ \frac{1}{\sqrt{4\pi t}} \left[\sum_{\tilde{\mathbf{q}}} \frac{-c(\tilde{\mathbf{q}}) \log \tilde{\mathbf{q}}}{\tilde{\mathbf{q}}^{\sigma_0/2}} e^{-(\log \tilde{\mathbf{q}})^2/4t} \right]$$

$$+ E_\Phi(t),$$

where

$$E_\Phi(t) = \frac{1}{2\pi} \int\limits_{-\infty}^{\infty} e^{-r^2 t} \Phi'/\Phi(\sigma_0/2 + ir)dr.$$

Remark 1. As in Chapter III, sums over ρ, α and the integral for E_Φ are to be understood as limits of sums and integral from $-T_m$ to T_m, taken over a suitably defined sequence $\{T_m\}$, depending on Z and Φ.

Remark 2. If $\Phi(s) = G(s)/\widetilde{G}(\sigma_0 - s)$, then

$$E_\Phi(r) = E_G(r) + E_{\widetilde{G}}(r)$$

where

$$E_G(r) = \frac{1}{2\pi} \int\limits_{-\infty}^{\infty} e^{-r^2 t} G'/G(\sigma_0/2 + ir)dr.$$

In particular, if $G = \widetilde{G}$, then $E_\Phi(r) = 2E_G(r)$.

Let us establish some notation in order to write the formula in Theorem 1.1 in a form fitting **AS 1**, **AS 2** and **AS 3**. Let $L = \{\mu_k\}$ be the set of numbers

$$L = \{\mu_k\} = \{-(\rho - \sigma_0/2)^2, -(\alpha - \sigma_0/2)^2\},$$

ordered as a sequence, with integral multiplicities

$$\{a(\mu_k)\} = \{v(\rho), v(\alpha)\}.$$

Define the associated theta function

$$(3) \qquad \theta_L(t) = \sum_\rho v(\rho)e^{(\rho - \sigma_0/2)^2 t} + \sum_\alpha v(\alpha)e^{(\alpha - \sigma_0/2)^2 t}.$$

Similarly, let $L^\vee$ be the family of numbers

$$L^\vee = \{\mu_k^\vee\} = \{(\log \mathbf{q})^2/4, (\log \widetilde{\mathbf{q}})^2/4\}$$

with (not necessarily integral) "multiplicities"

$$\{a(\mu_k^\vee)\} = \left\{ \frac{-c(\mathbf{q})\log \mathbf{q}}{\mathbf{q}^{\sigma_0/2}}, \frac{-c(\widetilde{\mathbf{q}})\log \widetilde{\mathbf{q}}}{\widetilde{\mathbf{q}}^{\sigma_0/2}} \right\},$$

and define the associated theta function

$$\theta_{L^\vee}(t) = \sum_{\mathbf{q}} \frac{-c(\mathbf{q}) \log \mathbf{q}}{\mathbf{q}^{\sigma_0/2}} e^{-[(\log \mathbf{q})^2/4]t}$$

(4)
$$+ \sum_{\tilde{\mathbf{q}}} \frac{-c(\tilde{\mathbf{q}}) \log \tilde{\mathbf{q}}}{\tilde{\mathbf{q}}^{\sigma_0/2}} e^{-[(\log \tilde{\mathbf{q}})^2/4]t}.$$

With this notation, we can now write Theorem 1.1 in the following form.

Theorem 1.2. *With notation as above, we have the inversion formula*

$$\theta_L(t) = \frac{1}{\sqrt{4\pi t}} \theta_{L^\vee}(1/t) + E_\Phi(t).$$

In general, we define a theta inversion formula to be a relation between two theta functions, such as (3) and (4), with an additional term such as $E_\Phi(t)$ which we require to satisfy the asymptotic condition **AS 2**.

Theorem 1.3. *With notation as above, the theta function*

$$\sum_\rho v(\rho) e^{(\rho-\sigma_0/2)^2 t}$$

*satisfies the asymptotic conditions **AS 1**, **AS 2**, and **AS 3**.*

Proof. By the Cramér theorem, we have that the sequences

$$\{\rho/i : Z(\rho) = 0 \text{ with } \operatorname{Im}(\rho) > 0 \text{ and } -a < \operatorname{Re}(\rho) < \sigma_0 + a\}$$

and

$$\{\rho/i : Z(\rho) = 0 \text{ with } \operatorname{Im}(\rho) < 0 \text{ and } -a < \operatorname{Re}(\rho) < \sigma_0 + a\}$$

are such that the associated theta functions satisfy the three basic asymptotic conditions. Therefore, by applying Theorem 7.6 and Theorem 7.7 of [JoL 93a], we conclude that the theta function

$$\sum_\rho v(\rho) e^{(\rho-\sigma_0/2)^2 t}$$

satisfies the asymptotic conditions. ☐

The theta series taken over $\{\alpha\}$ is incomplete, and it will be more useful in this chapter to deal with the alternate version of the explicit formula given as Theorem 2.3 in §2 of Chapter III.

We let R be of regularized harmonic series type. For each real number u such that R has no pole on the line $\text{Re}(z) = u$, we have the E_u-**transform**

$$E_u R(t) = \lim_{n \to \infty} \frac{1}{2\pi i} \int\limits_{u-iT_n}^{u+iT_n} e^{(z-\sigma_0/2)^2 t} R(z) dz$$

where $\{T_n\}$ is a sequence selected as in Theorem 6.2 of Chapter I.

Directly from Theorem 2.3 of Chapter III, we have the following result.

Theorem 1.4. *Let* $(Z, \tilde{Z}, \Phi)$ *be in the fundamental class. Decompose* Φ *as a product*

$$\Phi = \Phi_{\text{left}} \Phi_{\text{right}}$$

such that Φ_{left} *has all its zeros and poles in a left half plane and* Φ_{right} *has all its zeros and poles in a right half plane. Select* $A > \sigma_0$, $\delta > 0$, *and* a *such that:*

   Φ_{left} *has all its zeros and poles in* $\text{Re}(z) \le A - \delta$;
   Φ_{right} *has all its zeros and poles in* $\text{Re}(z) \ge -a + \delta$.

Let

$$\{\zeta\} = \text{set of zeros and poles of } \Phi_{\text{left}} \text{ with } \text{Re}(\zeta) > -a.$$

Then for all $t > 0$ *we have*

$$\sum_{\rho} v(\rho) e^{(\rho-\sigma_0/2)^2 t} + \sum_{\zeta} v(\zeta) e^{(\zeta-\sigma_0/2)^2 t}$$

$$= \frac{1}{\sqrt{4\pi t}} \left[\sum_{\mathbf{q}} \frac{-c(\mathbf{q}) \log \mathbf{q}}{\mathbf{q}^{\sigma_0/2}} e^{-(\log \mathbf{q})^2/4t} \right]$$

$$+ \frac{1}{\sqrt{4\pi t}} \left[\sum_{\tilde{\mathbf{q}}} \frac{-c(\tilde{\mathbf{q}}) \log \tilde{\mathbf{q}}}{\tilde{\mathbf{q}}^{\sigma_0/2}} e^{-(\log \tilde{\mathbf{q}})^2/4t} \right]$$

$$+ E_A(\Phi'_{\text{left}}/\Phi_{\text{left}})(t) + E_{-a}(\Phi'_{\text{right}}/\Phi_{\text{right}})(t).$$

We note that in Theorem 1.4 both theta series satisfy the asymptotic conditions **AS 1**, **AS 2**, and **AS 3**. This assertion is true for the sum over ζ since we assumed Φ was of regularized product type, because we choose $-a$ sufficiently far to the left, and so we can apply Theorem 7.6 and Theorem 7.7 of [JoL 93a]. As for the sum over ρ, one applies Cramér's theorem and Theorem 7.6 and Theorem. 7.7 of [JoL 93a].

Remark 3. In the present application, and the subsequent ones in this chapter and the next, we apply the explicit formula to the simplest types of Gaussian test functions. Even for such test functions, one can give examples where instead of $(4\pi t)^{1/2}$ or $(4\pi t)^{3/2}$ (as in Chapter V, §4) one takes $(4\pi t)^{n/2}$ for an odd integer n, and in addition one also has an arbitrary polynomial as a coefficient. If one uses $(4\pi t)^{n/2}$ with an even integer n, then one gets Bessel series instead of Dirichlet series. All these cases deserve to be treated systematically, since they apply to several important situations of spectral analysis for classical manifolds, and we shall do so elsewhere. Here we selected the simplest cases to serve as examples.

§2. Some examples of theta inversions

To emphasize the significance of the theta inversion formula given in Theorem 1.1, let us now discuss a few specific applications of the theorem.

Example 1: The sine function. Let us write the sine function as

$$\sin(\pi i s) = \frac{e^{-\pi s} - e^{\pi s}}{2i} = -\frac{e^{\pi s}}{2i}\left(1 - e^{-2\pi s}\right) = G(s)Z(s)$$

where $G(s) = -e^{\pi s}/2i$ and $Z(s) = 1 - e^{-2\pi s}$. The fact that the sine function is odd trivially yields the functional equation

$$G(s)Z(s) = -G(-s)Z(-s),$$

so $\sigma_0 = 0$. Further, Z satisfies the Euler sum condition since

$$\log Z(s) = \log\left(1 - e^{-2\pi s}\right) = \sum_{n=1}^{\infty} \frac{1}{n} e^{-2\pi n s},$$

whence $\{q\} = \{e^{2\pi n}\}$ for $n \geq 1$ and $c(e^{2\pi n}) = 1/n$. Since $\sin(\pi i s)$ is zero only when $s \in \mathbf{Z}$, and then with multiplicity one, the inversion formula Theorem 1.1 specializes to the equality

$$(1) \qquad \sum_{n=-\infty}^{\infty} e^{-n^2 t} = 2 \cdot \frac{1}{\sqrt{4\pi t}} \sum_{n=1}^{\infty} 2\pi e^{-(2\pi n)^2/4t} + E_\Phi(t),$$

where the factor of 2 appears since the sums over $\{q\}$ and over $\{\tilde{q}\}$ coincide in this example. Also, we have

$$(2) \qquad E_\Phi(t) = 2E_G(t) = \frac{1}{\pi} \int_{-\infty}^{\infty} e^{-r^2 t}\pi\, dr = \frac{\sqrt{\pi}}{\sqrt{t}}.$$

Combining the terms in (1) and (2), we obtain the classical Jacobi inversion formula, which is the relation

$$\frac{1}{2\pi} \sum_{n=-\infty}^{\infty} e^{-n^2 t} = \frac{1}{\sqrt{4\pi t}} \sum_{n=-\infty}^{\infty} e^{-(2\pi n)^2/4t} \qquad \text{for } t > 0.$$

Similarly, one can derive other classical formulas using the cosine function and hyperbolic trigonometric functions.

Example 2: Dirichlet polynomials. Recall the definition of a Dirichlet polynomial given in §7 of Chapter I. We saw in Chapter II that any Dirichlet polynomial P can be written as

$$P(s) = a_N b_N^s \left[1 + \sum_{n=1}^{N-1} \frac{a_n}{a_N} \left(\frac{b_n}{b_N} \right)^s \right] = a_N b_N^s \cdot Z(s),$$

where Z has an Euler sum and functional equation with fudge factor

$$\Phi(s) = \frac{a_N}{a_1} \left(\frac{b_N}{b_1} \right)^s.$$

Hence, one can directly evaluate the Weil functional, yielding the formula

$$E_\Phi(t) = \frac{1}{2\pi} \int_{-\infty}^{\infty} e^{-r^2 t} (\log(b_N/b_1)) dr = \frac{\log(b_N/b_1)}{\sqrt{4\pi t}}.$$

Therefore, the associated theta inversion formula is simply

$$\sum_\rho v(\rho) e^{\rho^2 t} = \frac{\log(b_N/b_1)}{\sqrt{4\pi t}} - \frac{1}{\sqrt{4\pi t}} \sum_q c(q)(\log q) e^{-(\log q)^2/4t}$$

$$- \frac{1}{\sqrt{4\pi t}} \sum_{\tilde{q}} c(\tilde{q})(\log \tilde{q}) e^{-(\log \tilde{q})^2/4t}.$$

The specific case of $N = 2$ with

$$b_2 = b_1^{-1} = e^{\pi s} \quad \text{and} \quad a_2 = -a_1 = i/2$$

yields the Jacobi inversion formula.

We find this example particularly interesting for the following reason. In Example 1, the Jacobi inversion formula is a relation involving two well-known sets of data, meaning that both theta functions in the inversion formula involves the squares of integers. However, in the case of a Dirichlet polynomial, we have one well-known set of numbers, namely the sets $\{q\}$ and $\{\tilde{q}\}$ which are explicitly and simply expressed in terms of the initial set of numbers $\{a_j, b_j\}$, and one unknown set of numbers, namely the set of zeros.

Example 3: Zeta functions of number fields. In §6 of Chapter III we gave an evaluation of the Weil functional associated to the classical gamma function. Hence, there is a theta inversion formula associated to any zeta function of a number field as in Theorem 1.1. The term E_Φ or E_G is given explicitly as follows. If

$$G(s) = A^{s/2}\Gamma(s)^{r_2}\Gamma(s/2)^{r_1},$$

then, by page 146 of [Ba 81], see also §4 of [JoL 93b], we find

$$E_G(t) = \frac{\log A^{1/2}}{2\pi} \int_{-\infty}^{\infty} e^{-r^2 t} dr + \frac{r_2}{2\pi} \int_{-\infty}^{\infty} \Gamma'/\Gamma(1/2 + it)e^{-r^2 t} dr$$

$$+ \frac{r_1/2}{2\pi} \int_{-\infty}^{\infty} \Gamma'/\Gamma(1/4 + it/2)e^{-r^2 t} dr$$

$$= \frac{\log A^{1/2}}{\sqrt{4\pi t}} + \frac{r_2}{\sqrt{4\pi t}} \int_{0}^{\infty} \left[\frac{1}{x} - e^{-x^2/4t}e^{x/2}\theta(x) \right] e^{-x} dx$$

$$+ \frac{r_1/2}{\sqrt{4\pi t}} \int_{0}^{\infty} \left[\frac{1}{x} - 2e^{-x^2/4t}e^{3x/2}\theta(2x) \right] e^{-2x} dx$$

where

$$\theta(x) = \sum_{n=0}^{\infty} e^{-nx} = \frac{1}{1 - e^{-x}}.$$

It is important to note that to zeta functions, we are associating theta functions which admit inversion formulas but are different from the theta functions used in Hecke's proof of the functional equation and meromorphic continuation (see, for example, Chapter XIII of [La 70]).

Example 4: A connection with regularized products. In spectral theory, one meets the situation when a certain operator has the sequence of eigenvalues $L = \{\lambda_k\}$ with integer multiplicities $\{a_k\}$. In such a situation, when the theory of regularized products applies (namely our axioms from [JoL 93a], recalled in Chapter I, §1), one may have the additional relation

$$D_L(s(s - \sigma_0)) = Z(s)G(s).$$

This occurs for instance for the Selberg zeta function and its analogues. Then

$$(\rho_k - \sigma_0/2)^2 = -\lambda_k + \sigma_0^2/4.$$

Thus we have the simple relation

$$\sum_k v(\rho_k)e^{(\rho_k - \sigma_0/2)^2 t} = e^{(\sigma_0^2/4)t} \sum_k a_k e^{-\lambda_k t} = e^{(\sigma_0^2/4)t}\theta_L(t),$$

expressing the theta series formed with the squares $(\rho_k - \sigma_0/2)^2$ in terms of the theta series with the eigenvalues.

As stated above, in the case of the operator d^2/dx^2 on the circle $S^1 = \mathbf{R}/2\pi\mathbf{Z}$, this relation reduces to the Jacobi theta series formed with $\lambda_k = k^2$ and $\sigma_0 = 0$. In both cases we are dealing with the eigenvalues of the positive Laplacian on some Riemannian manifold.

CHAPTER V

From Theta Inversions to Functional Equations

In this chapter we shall carry out the inverse construction of the preceeding sections; that is, from a theta series satisfying an inversion formula, we derive, by means of a Gauss transform, a Dirichlet series satisfying an additive functional equation.

Of course, Riemann's proof of the functional equation of the Riemann zeta function also relied on a theta inversion formula, but in the present situation, our use of theta inversion is different from that of Riemann because we take a Laplace transform, with a quadratic change of variables, of the regularized theta series instead of the Mellin transform of the theta series. Hence we construct new types of zeta functions which are essentially regularized harmonic series. In case the residues are integers, such series are logarithmic derivatives of functions in the fundamental class. Thus, we see that the general theory requires that we consider the additive class rather than the more restrictive multiplicative class.

Let R be of regularized harmonic series type. In §1, we analyze the transform E_u defined by

$$E_u R(t) = \lim_{n \to \infty} \frac{1}{2\pi i} \int\limits_{u - iT_n}^{u + iT_n} e^{(z - \sigma_0/2)^2 t} R(z) dz.$$

In §2 we carry out the properties of the Gauss transform of theta series, inverted theta series, and the transform $E_u R$. These properties are shown to imply the functional equation of the new zeta function in §3. In §4 we work out an example to obtain a zeta function for any compact quotient of the three dimensional hyperbolic space $\mathbf{M}_3$. This zeta function lies in our additive class but, in general, not in our multiplicative class. For extensions of this example, see the remarks at the end of §4.

Throughout the remainder of this chapter, unless otherwise specified, we use the following basic conditions.

The basic conditions

We let σ_0 be a real number ≥ 0.

We let R be a function of regularized harmonic series type.

We let $\{\mu_k\}$ and $\{a_k\}$ satisfy **DIR 1**, **DIR 2**, and **DIR 3**, and we assume that the corresponding theta series

$$\theta(t) = \theta_L(t) = \sum_{k=0}^{\infty} a_k e^{-\mu_k t}$$

satisfies **AS 1**, **AS 2**, and **AS 3**. We let (M_θ, m) be the reduced order of the theta series θ. Note that we have used μ_k instead of λ_k in the notation. This is because in a subsequent application to spectral theory, there will be an operator with eigenvalues λ_k which are translates of μ_k, namely

$$\mu_k = \lambda_k - \sigma_0^2/4.$$

We let R_θ be the regularized harmonic series associated to the theta function θ, as defined in Chapter I, §1; that is,

$$R_\theta(z) = \mathrm{CT}_{s=1} \mathbf{LM}\theta(s, z).$$

We let $\{\mathbf{q}\}$ be a sequence of real numbers > 1 converging to infinity. We let $\{c(\mathbf{q})\}$ be a sequence of complex numbers, and we let

$$Л(s) = \sum_{\mathbf{q}} \frac{c(\mathbf{q})}{\mathbf{q}^s}$$

be the associated Dirichlet series, which we assume converges absolutely in some right half plane. Thus,

$$Л'(s) = \sum_{\mathbf{q}} \frac{-c(\mathbf{q}) \log \mathbf{q}}{\mathbf{q}^s}.$$

We let

$$\theta_{L^\vee}(t) = \sum_{\mathbf{q}} \frac{-c(\mathbf{q}) \log \mathbf{q}}{\mathbf{q}^{\sigma_0/2}} e^{-[(\log \mathbf{q})^2/4]t}.$$

§1. The Weil functional of a Gaussian test function.

In this section, we are concerned with a function R of regularized harmonic series type. For each $\zeta \in \mathbf{C}$ we let:

$$a_R(\zeta) = a(\zeta) = \text{ residue of } R \text{ at } \zeta.$$

Observe that if $R = R_\Phi = \Phi'/\Phi$, where Φ is of regularized product type, then $a(\zeta) = v(\zeta)$ is the order of Φ at ζ.

Any function R of regularized series type can be expressed as a sum of two such functions, each of which has poles only in a half plane to the left or a half plane to the right. Let us write such a decomposition as

$$R(z) = R_{\text{left}}(z) + R_{\text{right}}(z).$$

Assume that $A > 0$ is chosen so that all the poles of R_{left} lie in a left half plane of the form $\text{Re}(z) \leq A - \delta$, for some $\delta > 0$. Similarly, assume that $a > 0$ is chosen so that all the poles of R_{right} lie in a right half plane of the form $\text{Re}(z) \geq -a + \delta$, for some $\delta > 0$. Assume that R_{left} has reduced order $(M_{\text{left}}, m_{\text{left}})$ and R_{right} has reduced order $(M_{\text{right}}, m_{\text{right}})$

As a direct application of Theorem 6.2(a) of Chapter I, we obtain the following result.

Lemma 1.1. *With notation as above, we have*

$$E_A R_{\text{left}}(t) = O(t^{-(M_{\text{left}}+2)/2}) \qquad \text{for } t \to 0$$

and

$$E_{-a} R_{\text{right}}(t) = O(t^{-(M_{\text{right}}+2)/2}) \qquad \text{for } t \to 0.$$

Proof. By Theorem 6.2(a) of Chapter I, we know that on the line $\text{Re}(z) = A$ the function R_{left} has polynomial growth. In fact, the integral giving $E_A R_{\text{left}}$ can be coarsely estimated by

$$O\left(\int_{-\infty}^{\infty} e^{-btu^2} |u|^{M_{\text{left}}+2} \frac{du}{u} \right)$$

with some number $b > 0$. The first estimate asserted then follows from the standard change of variables $y = \sqrt{t}u$. The second estimate is proved similarly. $\square$

The proof of Lemma 1.1 also applies to prove the following asymptotic bounds.

Lemma 1.2. *With notation as above, we have*

$$E_A R_{\text{left}}(t) = O(t^{-(M_{\text{left}}+2)/2}) \qquad \text{for } t \to \infty$$

and

$$E_{-a} R_{\text{right}}(t) = O(t^{-(M_{\text{right}}+2)/2}) \qquad \text{for } t \to \infty.$$

Note that the bound given in Theorem 6.2(a) of Chapter I is stronger than what is used in the proof of Lemma 1.1 and Lemma 1.2 above. However, the above results are sufficient for our purposes. In fact, one can easily improve the exponent to $-(M+1)/2 + \epsilon$ for any $\epsilon > 0$.

The bounds in Lemma 1.1 and Lemma 1.2 are enough to allow us to deal with the Gauss transforms of $E_A R_{\text{left}}$ and $E_{-a} R_{\text{right}}$ in the next section.

§2. Gauss transforms.

We shall need some analysis concerning Bessel integrals and what we call the Gauss transform, which we carry out in this section. Throughout we let r, t, u, x be real variables with $t, u, x > 0$. We start by recalling the collapse of the Bessel integral $K_s(x, u)$ under certain conditions.

Lemma 2.1. *Let*

$$K_s(x, u) = \int_0^\infty e^{-x^2 t} e^{-u^2/t} t^s \frac{dt}{t}.$$

Then:

a) *In the case $s = 1/2$, we have the evaluation*

$$K_{1/2}(x, u/2) = \frac{\sqrt{\pi}}{x} e^{-ux}.$$

b) *Let $Д_x$ and $Д_u$ be the differential operators*

$$Д_x = -\frac{1}{2x} \frac{\partial}{\partial x} \quad \text{and} \quad Д_u = -\frac{1}{2u} \frac{\partial}{\partial u}.$$

Then we have the relations

$$(Д_x)^n K_s(x, u) = K_{s+n}(x, u)$$

and

$$(Д_u)^n K_s(x, u) = K_{s-n}(x, u).$$

For further properties of the Bessel integral, including a proof of Lemma 2.1, the reader is referred to [La 87].

Warning. For our purposes, we use a normalization of the Bessel integral slightly different from the classical one. If, for instance, $K_s^B(c)$ denotes the K-Bessel function which one finds in tables (see Magnus, Oberhettinger, Erdelyi, Whittaker and Watson, etc.), then we have the relation

$$2K_s^B(2c) = K_s(c).$$

where

$$K_s(c) = \int\limits_0^\infty e^{-c(t+1/t)} t^s \frac{dt}{t}.$$

The classical normalization gets rid of some factors in the differential equation, and our normalization gets rid of extraneous factors in the above integral.

For a suitable function f of a single real variable, we recall the Laplace-Mellin transform is defined as

$$\mathbf{LM}f(s,z) = \int\limits_0^\infty f(t)e^{-zt} t^s \frac{dt}{t}.$$

We shall deal with $f = \theta$, where θ is a theta series of the sort considered previously. We shall put $s = N + 1$, where N is an integer ≥ 0, sufficiently large to cancel the singularity at 0, and we shall also make a change of variables with $z = (s - \sigma_0/2)^2$. With this, we obtain an integral operator which we call the **Gauss transform**.

More precisely, let N be a positive integer and let σ_0 be a real number ≥ 0. For a suitable test function f, we define

$$\mathrm{Gauss}^{(N)}_{\sigma_0/2}(f)(s) = (2s - \sigma_0) \int\limits_0^\infty f(t) e^{-(s-\sigma_0/2)^2 t} t^{N+1} \frac{dt}{t}.$$

Then

$$\mathrm{Gauss}^{(N)}_{\sigma_0/2}(f)(s) = \frac{dz}{ds} \cdot \mathbf{LM}f(N+1, (s - \sigma_0/2)^2).$$

Because of the change of variables from z to s, we are led to consider the differential operator

$$\mathcal{D}_{s-\sigma_0/2} = -\frac{d}{ds} \circ \frac{1}{2s - \sigma_0}.$$

Let $x = s - \sigma_0/2$. For every integer $N \geq 0$, we then get the formula

$$2x \cdot \mathit{Д}_x^N \circ \frac{1}{2x} = \mathcal{D}^N_{s-\sigma_0/2}.$$

Gauss transform of an inverted theta series

We now shall prove inversion formulas for Gauss transforms of certain series. Such series occurred in the previous section in connection with functions in the fundamental class. Here we start with series for which we assume only the basic conditions.

Directly from Lemma 2.1, we have the following theorem.

Theorem 2.2. *Let*

$$G(t) = \frac{1}{\sqrt{4\pi t}} \sum_{\mathbf{q}} \frac{-c(\mathbf{q}) \log \mathbf{q}}{\mathbf{q}^{\sigma_0/2}} e^{-(\log \mathbf{q})^2/4t}.$$

Then for s real and sufficiently large, and $N \geq 0$, we have

$$\mathrm{Gauss}^{(N)}_{\sigma_0/2}(G)(s) = \mathcal{D}^N_{s-\sigma_0/2} \mathcal{Л}'(s).$$

Proof. The proof follows immediately from the definitions and Lemma 2.1. For each $\mathbf{q}$, we put $x = s - \sigma_0/2$ and $u = \log \mathbf{q}$ and compute the transform of each individual term

$$g_{\mathbf{q}}(t) = \frac{1}{\sqrt{4\pi t}} e^{-u^2/4t}$$

as follows:

$$\mathrm{Gauss}^{(N)}_{\sigma_0/2}(g_{\mathbf{q}})(s) = \frac{2x}{\sqrt{4\pi}} K_{N+1/2}(x, u/2)$$

$$= \frac{2x}{2\sqrt{\pi}} \mathcal{D}^N_x K_{1/2}(x, u/2)$$

$$= \frac{2x}{2\sqrt{\pi}} \mathcal{D}^N_x \left(\frac{\sqrt{\pi}}{x} e^{-ux} \right)$$

$$= 2x \mathcal{D}^N_x \left(\frac{1}{2x} e^{-ux} \right)$$

$$= \mathcal{D}^N_x e^{-(\log \mathbf{q})x}$$

$$= \mathcal{D}^N_{s-\sigma_0/2} \left(\mathbf{q}^{-s} \mathbf{q}^{\sigma_0/2} \right),$$

from which the formula in Theorem 2.2 is now clear. $\square$

Gauss transform of a theta series

Next, let us apply the Gauss transform to a theta series itself; that is, let us evaluate

$$\text{Gauss}^{(N)}_{\sigma_0/2}(\theta)(s) = (2s - \sigma_0) \int_0^\infty \theta(t) e^{-(s-\sigma_0/2)^2 t} t^{N+1} \frac{dt}{t}$$

where

$$\theta(t) = \sum_{k=0}^\infty a_k e^{-\mu_k t}.$$

Let ρ_k be complex numbers, all but a finite number of which have positive imaginary part, and such that

$$(\rho_k - \sigma_0/2)^2 = -\mu_k.$$

Theorem 2.3. *For any positive integer N sufficiently large, and s real and sufficiently large, we have*

$$\int_0^\infty \theta(t) e^{-(s-\sigma_0/2)^2 t} t^{N+1} \frac{dt}{t} = \sum_{k=0}^\infty a_k \frac{\Gamma(N+1)}{[(s-\rho_k)(s+\rho_k-\sigma_0)]^{N+1}}$$

$$= \sum_{k=0}^\infty a_k \frac{\Gamma(N+1)}{[(s-\sigma_0/2)^2 - (\rho_k-\sigma_0/2)^2]^{N+1}},$$

hence the series gives the meromorphic continuation of the integral to all $s \in \mathbf{C}$ with poles at the points $s = \rho_k$ and $s = \sigma_0 - \rho_k$.

Proof. This identity follows routinely by integrating term by term, using the change of variables $t \mapsto t/(s - \sigma_0/2)^{1/2}$. There is no problem with the square root since s was assume to be real and sufficiently large. $\square$

Next we obtain the inversion analogous to Theorem 2.2, but for the theta series instead of the **q** series.

Theorem 2.4. *For any positive integer N sufficiently large, and s real and sufficiently large, we have*

$$\text{Gauss}^{(N)}_{\sigma_0/2}(\theta)(s) = \mathcal{D}^{(N)}_{s-\sigma_0/2}\left[(2s-\sigma_0)R_\theta((s-\sigma_0/2)^2)\right].$$

Proof. As a function of s, $(2s-\sigma_0)R_\theta((s-\sigma_0/2)^2)$ is a meromorphic function whose singularities are simple poles at the points $s = \rho_k$ and $s = \sigma_0 - \rho_k$ with residue a_k. For any individual k, we have

$$\left(-\frac{\partial}{\partial z}\right)^N\left(\frac{1}{z+\mu_k}\right) = \frac{\Gamma(N+1)}{(z+\mu_k)^{N+1}}.$$

So,

$$\left(\frac{-1}{2s-\sigma_0}\frac{\partial}{\partial s}\right)^N\left[\frac{1}{(s-\sigma_0/2)^2-(\rho_k-\sigma_0/2)^2}\right]$$
$$= \frac{\Gamma(N+1)}{[(s-\sigma_0/2)^2-(\rho_k-\sigma_0/2)^2]^{N+1}},$$

which can be written as

$$= \frac{1}{2s-\sigma_0}\mathcal{D}^N_{s-\sigma_0/2}\left[\frac{2s-\sigma_0}{(s-\sigma_0/2)^2-(\rho_k-\sigma_0/2)^2}\right].$$

The result now follows from Theorem 2.3 after multiplying by the term $2s-\sigma_0$ and then summing over k, which is valid since for N sufficiently large the series is absolutely convergent. $\square$

Remark 1. From Theorem 7.6 and Theorem 7.7 of [JoL 93a], one can show that it suffices to take $N > (M_\theta + 2)/2$, where θ has reduced order (M_θ, m). In fact, this requirement can be easily improved to $N > (M_\theta + 1)/2 + \epsilon$ for any $\epsilon > 0$.

Remark 2. In the degenerate case when $\rho_k = \sigma_0/2$, the regularized harmonic series with the quadratic change of variables has a double pole at $s = \sigma_0/2$ with zero residue, which is consistent with the above evaluation of residues.

Inverse Gauss transform of a regularized harmonic series type

Finally, we deal with the Gauss transform applied to the terms $E_A R_{\text{left}} \ E_{-a} R_{\text{right}}$ which arose in Chapter IV. The next lemma evaluates this Gauss transform, and amounts to an inversion formula, showing that up to taking derivatives, the Gauss transform is the inverse of the E_A transform.

Theorem 2.5. *With notation as above, we have, for N sufficiently large and s real and sufficiently large, the equalities*

$$\text{Gauss}^{(N)}_{\sigma_0/2}(E_A R_{\text{left}})(s) = \mathcal{D}^N_{s-\sigma_0/2} R_{\text{left}}(s)$$

and

$$\text{Gauss}^{(N)}_{\sigma_0/2}(E_{-a} R_{\text{right}})(s) = \mathcal{D}^N_{s-\sigma_0/2} R_{\text{right}}(\sigma_0 - s).$$

Proof. Let

$$F(s) = \int\limits_0^\infty E_A R_{\text{left}}(t) e^{-(s-\sigma_0/2)^2 t} t^{N+1} \frac{dt}{t}$$

which is defined for N sufficiently large by Lemma 1.1, and s real and sufficiently large by Lemma 1.2. Then

$$(1) \quad F(s) = \frac{1}{2\pi i} \int\limits_0^\infty \int\limits_{A-i\infty}^{A+i\infty} e^{(z-\sigma_0/2)^2 t} e^{-(s-\sigma_0/2)^2 t} R_{\text{left}}(z) t^{N+1} dz \frac{dt}{t}.$$

By changing the order of integration, which is valid by Theorem 6.2(a) of Chapter I, we may write (1) in the form

$$(2) \quad F(s) = \frac{1}{2\pi i} \int\limits_{A-i\infty}^{A+i\infty} \frac{\Gamma(N+1)}{[(s-\sigma_0/2)^2 - (z-\sigma_0/2)^2]^{N+1}} R_{\text{left}}(z) dz.$$

Let us write

$$\frac{\Gamma(N+1)}{[(s-\sigma_0/2)^2 - (z-\sigma_0/2)^2]^{N+1}}$$

$$= \Pi^N_{s-\sigma_0/2} \left[\frac{1}{(s-\sigma_0/2)^2 - (z-\sigma_0/2)^2} \right],$$

which allows us to express (2) as

$$F(s)$$

(3)
$$= \frac{1}{2\pi i} \int_{A-i\infty}^{A+i\infty} \amalg_{s-\sigma_0/2}^{N} \left[\frac{1}{(s-\sigma_0/2)^2 - (z-\sigma_0/2)^2} \right] R_{\text{left}}(z)dz.$$

Because of the symmetry in s and z, we can write (3), as

$$F(s)$$

$$= \frac{1}{2\pi i} \int_{A-i\infty}^{A+i\infty} (\amalg_{z-\sigma_0/2})^N \left[\frac{(-1)^N}{(s-\sigma_0/2)^2 - (z-\sigma_0/2)^2} \right] R_{\text{left}}(z)dz$$

$$= \frac{1}{2\pi i} \int_{A-i\infty}^{A+i\infty} (-\amalg_{z-\sigma_0/2})^N \left[\frac{1}{(s-z)(s+z-\sigma_0)} \right] R_{\text{left}}(z)dz.$$

Integration by parts shows that $-\amalg_{z-\sigma_0/2}$ and $\mathcal{D}_{z-\sigma_0/2}$ are adjoint operators. Hence, we obtain the formula

$$\text{Gauss}_{\sigma_0/2}^{(N)}(E_A R_{\text{left}})(s) = (2s - \sigma_0)F(s)$$

$$= \frac{1}{2\pi i} \int_{A-i\infty}^{A+i\infty} \left[\frac{1}{s-z} + \frac{1}{s+z-\sigma_0} \right] (\mathcal{D}_{z-\sigma_0/2})^N R_{\text{left}}(z)dz.$$

From the *proof* of Theorem 6.2(a) of Chapter I, one obtains that on the line $\text{Re}(z) = A$ we have

(4) $(\mathcal{D}_{z-\sigma_0/2})^k R_{\text{left}}(A+it) = O(|t|^{M+\epsilon-2k})$ for $t \to \infty$,

for any $\epsilon > 0$ and positive integer k. So, for $N > (M+1)/2 + \epsilon$, the integration by parts is valid.

Now we can evaluate $F(s)$ using a contour involving a half-circle, opening to the right with diameter along the vertical line $\text{Re}(s) = A$. In this region, $R_{\text{left}}(z)$ is holomorphic, so the only pole of the integrand occurs when $z = s$, and that with residue 1, after correcting the orientation of the contour. The integral along the half-circle will approach zero as the radius approaches infinity since, again by the *proof* of Theorem 6.2(a) of Chapter I, since the function $(\mathcal{D}_{z-\sigma_0/2})^k R_{\text{left}}(z)$ has polynomial growth as in (4). With all this, the first part of Theorem 2.5 is proved.

Concerning the second part of Theorem 2.5, one applies the same argument as above, except using a contour integral opening to the left, together with the identity $\mathcal{D}_z = \mathcal{D}_{-z}$. $\square$

Remark 2. Lemma 1.1 and the estimate in (4) show that it suffices to take $N > (M+1)/2 + \epsilon$ where R has reduced order (M, m).

§3. Theta inversions yield zeta functions

We now put the results of §2 together to show how theta inversion gives rise to an additive functional equation.

Given R of regularized harmonic series type, there exist a and A sufficiently large positive so that we can decompose R into a sum

$$R = R_{\text{left}} + R_{\text{right}}$$

where R_{left} has it poles in a left half plane $\operatorname{Re}(z) \leq A - \delta$, and where R_{right} has it poles in a right half plane $\operatorname{Re}(z) \geq -a + \delta$.

Theorem 3.1. *Assume that the theta series satisfies the inversion formula*

$$\theta_L(t) = \frac{1}{\sqrt{4\pi t}} \theta_{L^\vee}(1/t) + E_A R_{\text{left}}(t) + E_{-a} R_{\text{right}}(t).$$

For N sufficiently large, we have

$$\mathcal{D}^N_{s-\sigma_0/2}\left[(2s - \sigma_0)R_{\theta_L}((s - \sigma_0/2)^2)\right]$$
$$= \mathcal{D}^N_{s-\sigma_0/2}\left[\text{\Cyrillic{J}I}'(s) + R_{\text{left}}(s) + R_{\text{right}}(\sigma_0 - s)\right].$$

Proof. We simply apply Theorem 2.2, Theorem 2.3, and Theorem 2.5. $\square$

Remark 1. It suffices to take

$$N > \frac{1}{2}(\max\{M_{R_{\text{left}}}, M_{R_{\text{right}}}, M_\theta\} + 1) + \epsilon,$$

for any $\epsilon > 0$.

Theorem 3.2. *There is a polynomial P' of degree $\leq 2N - 1$ such that for all real s sufficiently large, we have*

$$(2s - \sigma_0)R_{\theta_L}((s - \sigma_0/2)^2)$$
$$= \text{Л}'(s) + R_{\text{left}}(s) + R_{\text{right}}(\sigma_0 - s) + P'(s).$$

Proof. Let

$$F(s) = (2s - \sigma_0)R_{\theta_L}((s - \sigma_0/2)^2) - \text{Л}'(s)$$
$$- R_{\text{left}}(s) - R_{\text{right}}(\sigma_0 - s),$$

and let us write the formula in Theorem 3.1 as

$$\left(-\frac{\partial}{\partial s}\frac{1}{2s - \sigma_0}\right)\mathcal{D}^{N-1}_{s-\sigma_0/2}F(s) = 0.$$

Hence, there exists a polynomial P_1 of degree ≤ 1 such that

$$\mathcal{D}^{N-1}_{s-\sigma_0/2}F(s) = P_1(s).$$

The proof finishes by continuing to unwind the differential operator $\mathcal{D}^{N-1}_{s-\sigma_0/2}$. $\square$

Corollary 3.3. *The Dirichlet series $\text{Л}'(s)$ has a meromorphic continuation such that under that map $s \mapsto \sigma_0 - s$ the function*

$$\text{Л}'(s) + R_{\text{left}}(s) + R_{\text{right}}(\sigma_0 - s) + P'(s)$$

is odd.

This is immediate from Theorem 3.2 and the formula in Theorem 2.3. Upon integrating, the functional equation obtained in the present case is what we call an additive functional equation, to distinguish it from the multiplicative functional equation satisfied by functions in the fundamental class.

We have now concluded the process from which a theta function with an inversion formula leads to a Dirichlet series having an additive functional equation with fudge terms which are of regularized harmonic series type.

Example 1. If we apply the above construction to the Riemann zeta function, then we have

$$R(s) = \frac{1}{2}\Gamma'/\Gamma(s/2) + \frac{1}{2}\Gamma'/\Gamma((1-s)/2) - \log \pi,$$

hence we have

$$R_{\text{left}}(s) = \frac{1}{2}\Gamma'/\Gamma(s/2) - \log \pi,$$

and

$$R_{\text{right}}(s) = \frac{1}{2}\Gamma'/\Gamma((1-s)/2).$$

It is easy to see that $\Pi'(s) = 2\zeta'_{\mathbf{Q}}/\zeta_{\mathbf{Q}}(s)$ and $M = 0$, hence the above theorems apply with $N = 1$. Therefore, Corollary 3.3 reconstructs the fact that the function

$$2\xi'_{\mathbf{Q}}/\xi_{\mathbf{Q}}(s) = 2\zeta'_{\mathbf{Q}}/\zeta_{\mathbf{Q}}(s) + \Gamma'/\Gamma(s/2) - \log \pi$$

satisfies the functional equation $\xi'_{\mathbf{Q}}/\xi_{\mathbf{Q}}(s) = -\xi'_{\mathbf{Q}}/\xi_{\mathbf{Q}}(1-s)$.

Remark 2. It may seem odd that whereas we started with a triple $(Z, \tilde{Z}, \Phi)$ in our fundamental class with functional equation which is not symmetric, we derived a function in Corollary 3.3 which has a symmetric functional equation up to an additive polynomial factor. The reason for this is the following. Since the test function to which we apply Theorem 2.3 of Chapter III is even, we can not distinguish between elements in the set $\{\mathbf{q}\}$ and elements in the set $\{\tilde{\mathbf{q}}\}$. Therefore, we actually ended up considering the zeta function $H_1(s) = Z(s)\tilde{Z}(s)$ which satisfies the symmetric functional equation

$$H_1(s)\Phi(s) = H_1(\sigma_0 - s)\Phi(\sigma_0 - s).$$

One could equally well use the odd test function

$$F_t(x) = \frac{-x/2t}{\sqrt{4\pi t}} e^{-x^2/4t},$$

so then

$$\mathbf{M}_{\sigma_0/2}f(s) = -(s - \sigma_0/2)e^{(s-\sigma_0/2)^2 t}.$$

After carrying through the computations as above, we end up with a formula of the form

$$(2s - \sigma_0)R_\theta((s - \sigma_0/2)^2)$$
$$= \Pi'(s) - (s - \sigma_0/2)\left[R_{\text{left}}(s) - R_{\text{right}}(\sigma_0 - s)\right] + P'(s).$$

This equation corresponds to the functional equation

$$H_2(s)H_2(\sigma_0 - s) = 1$$

where

$$H_2(s) = \frac{Z(s)}{\tilde{Z}(s)}\Phi(s).$$

Then, combining the two functional equations above, we recover the original functional equation (up to a sign)

$$(Z(s)\Phi(s))^2 = (\tilde{Z}(\sigma_0 - s))^2,$$

in other words

$$Z(s)\Phi(s) = \pm\tilde{Z}(\sigma_0 - s).$$

Example 2. The functional equation for the function H_2 in Remark 2 is like that of the scattering determinant associated to any non-compact hyperbolic Riemann surface of finite volume (see [He 83]).

§4. A new zeta function for compact quotients of M_3.

In this section we will give, what we believe, is a new example of a zeta function. The approach is that of the previous sections, namely to every theta inversion formula there is a corresponding zeta function. As we will see, the "zeta function" has a Dirichlet series in a half-plane, but is a regularized harmonic series whose singularities are simple poles whose residues are not necessarily integers. Hence the function is not the logarithmic derivative of a regularized product type.

Let M_3 be the simply connected three dimensional Riemannian manifold whose metric has constant sectional curvature equal to -1; see page 38 of [Ch 84] for a precise model. On M_3 there is a transitive group of isometries, so the heat kernel relative to the Laplacian acting on smooth functions, which we denote by K_{M_3}, is a function of distance r and t. In fact, let

$$(1) \qquad h_3(r,t) = \frac{1}{(4\pi t)^{3/2}} e^{-r^2/4t} e^{-t} \frac{r}{\sinh r}.$$

Then

$$K_{M_3}(\tilde{x}, t, \tilde{y}) = h_3(d_{M_3}(\tilde{x}, \tilde{y}), t).$$

For details of this computation, see page 150 of [Ch 84] or page 397 of [DGM 76]. The result is basically due to Millson.

Let $X = \Gamma \backslash M_3$ be a compact quotient of M_3. Using the spectral decomposition of the heat kernel K_X on X, one has the expression

$$(2) \qquad K_X(x,t,y) = \sum_{k=0}^{\infty} \phi_k(x)\phi_k(y)e^{-\lambda_k t},$$

where $\{\phi_k\}$ is a complete system of orthonormal eigenfunctions of the Laplacian and $\{\lambda_k\}$ is the corresponding system of eigenvalues. Choose $\tilde{x}$ and $\tilde{y}$ in M_3 lying above x and y in X. Then

$$(3) \qquad K_X(x,t,y) = \sum_{\gamma \in \Gamma} K_{M_3}(\tilde{x}, t, \gamma\tilde{y}).$$

View the pair (x, y) as fixed, and for each $\gamma \in \Gamma$ define

$$\log \mathbf{q}_\gamma(x, y) = \log \mathbf{q}_\gamma = d_{M_3}(\tilde{x}, \gamma\tilde{y}).$$

Thus we obtain a sequence $\{\mathbf{q}_\gamma\}_{\gamma\in\Gamma}$, which we reindex by itself as simply $\{\mathbf{q}\}$. Suppose $x \neq y$. Then by (1), we have

$$(4) \qquad K_X(x,t,y) = \frac{2}{(4\pi t)^{3/2}} e^{-t} \sum_{\mathbf{q}} \frac{\log \mathbf{q}}{\mathbf{q}-\mathbf{q}^{-1}} e^{-(\log \mathbf{q})^2/4t},$$

which we call the Γ-**expansion** or the $\mathbf{q}$-**expansion**. Combining (2) and (4), we immediately obtain the theta inversion formula

$$
\begin{aligned}
(5) \qquad & \sum_{k=0}^{\infty} \phi_k(x)\phi_k(y)e^{-\lambda_k t} \\
&= \frac{2}{(4\pi t)^{3/2}} e^{-t} \sum_{\mathbf{q}} \frac{\log \mathbf{q}}{\mathbf{q}-\mathbf{q}^{-1}} e^{-(\log \mathbf{q})^2/4t}.
\end{aligned}
$$

Observe that the factor e^{-t} could be brought to the other side. In the notation of §3, $\sigma_0 = 2$ and $\mu_k = \lambda_k - 1$. We note that (5) holds for any points $x \neq y$ on X, and we have not taken a trace of the heat kernel, although we have taken the trace with respect to the infinite Galois group Γ, i.e. the fundamental group of X. Formula (5) simply reflects a combination of the existence and uniqueness of the heat kernel on X together with the spectral expression (2) and the group expression (3) for K_X.

As in the previous sections, any theta inversion formula can be used to obtain a zeta function with additive functional equation. Let us carry through the computations in this case. We are still assuming $x \neq y$. From (5), let us derive two expressions for the function

$$(6) \qquad F_{x,y}(s) = (2s-2) \int_0^{\infty} K_X(x,t,y)e^{-s(s-2)t}t^2\, dt.$$

Using the spectral expansion (2), the integral in (6) yields the equality

$$(7) \qquad F_{x,y}(s) = (2s-2) \sum_{k=0}^{\infty} \frac{2\phi_k(x)\phi_k(y)}{(s(s-2)+\lambda_k)^3},$$

which can be shown to converge uniformly and absolutely on X by combining standard asymptotic formulas for eigenvalues on compact manifolds (see [BGV 92], for example) and Sogge's theorem on sup-norm bounds for eigenfunctions (see page 226 of [St 90]) .

On the other hand, we shall also take the Gauss transform of the right side of (5), for which we need the collapse of the Bessel integral

$$K_{3/2}(s-1, \log q) = \int_0^\infty e^{-(s-1)^2 t} e^{-(\log q)^2/4t} t^{3/2} \frac{dt}{t}$$

$$= \frac{-1}{2s-2} \frac{\partial}{\partial s} \left(\frac{\sqrt{\pi}}{s-1} q^{-(s-1)} \right),$$

by Lemma 2.1. Therefore, taking the Gauss transform of (3), or the right side of (5), we obtain a Bessel sum for $F_{x,y}$ which collapses to a Dirichlet series, for $\mathrm{Re}[(s-1)^2] > 0$, namely

$$F_{x,y}(s) = \frac{2}{(4\pi)^{3/2}} \sum_q \frac{\log q}{q - q^{-1}} (2s-2) K_{3/2}(s-1, \log q)$$

which becomes

$$(8) \qquad F_{x,y}(s) = \frac{1}{2\pi} \frac{\partial}{\partial s} \left(\frac{-1}{2s-2} \sum_q \frac{\log q}{1-q^{-2}} q^{-s} \right).$$

Using a simple argument involving volumes of fundamental domains, one can show directly that the Dirichlet series (8) converges for $\mathrm{Re}(s) > 2$. Therefore, we have for s in this half-plane, the formula

$$F_{x,y}(s) = \frac{1}{2\pi} \frac{\partial}{\partial s} \left(\frac{-1}{2s-2} \sum_q \frac{\log q}{1-q^{-2}} q^{-s} \right)$$

$$= (2s-2) \sum_{k=0}^\infty \frac{2\phi_k(x)\phi_k(y)}{(s(s-2)+\lambda_k)^3}.$$

From (7) it follows that the Dirichlet series in (8) has a meromorphic continuation to all $s \in \mathbb{C}$ with the additive functional equation

$$F_{x,y}(s) = -F_{x,y}(2-s).$$

Further, the meromorphic continuation has singularities precisely when $s(s-2) = -\lambda_k$, that is

$$(9) \qquad s = 1 \pm i\sqrt{\lambda_k - 1}$$

each singularity being a double pole with zero residue. Upon integrating with respect to s, we obtain the Dirichlet series

$$(10) \qquad Л_{x,y}(s) = \sum_q \frac{1}{1 - q^{-2}} q^{-s} = \sum_q \sum_{k=0}^{\infty} q^{-s-2k},$$

which satisfies the relation

$$\mathcal{D}_{s-1} Л'_{x,y}(s) = -2\pi F_{x,y}(s).$$

The function $Л = Л_{x,y}$ plays the role of the logarithm of a zeta function in the fundamental class, and would be such a logarithm if the residues of $Л'$ were integers, but they are not in general.

Theorem 4.1. *The function $Л$ has the following properties:*

  i) *The series for $Л(s)$ converges for $\mathrm{Re}(s) > 2$;*
  ii) *$Л'$ has a meromorphic continuation to all $s \in \mathbf{C}$;*
  iii) *$Л'$ is odd under the map $s \mapsto 2 - s$;*
  iv) *The singularities of the meromorphic continuation of the function $(2s - 2)Л'(s)$ are all simple poles at the points (9), with corresponding residue $2\pi\phi_n(x)\phi_n(y)$.*

Property (iv) should be viewed as a type of Riemann hypothesis for the zeta function (10).

We conclude with some remarks in the case $x = y$. We pick $\tilde{x} = \tilde{y}$. Then the sum in (1) must be decomposed into the term with $\gamma = \mathrm{id}$ and the other terms. We have $q_{\mathrm{id}} = 1$ and $q_\gamma > 1$ for all $\gamma \neq \mathrm{id}$. In the present case, the term with $\gamma = \mathrm{id}$ is easily computed, and instead of (4) we then obtain
(11)

$$K_X(x,t,x) = \frac{1}{(4\pi t)^{3/2}} e^{-t} + \frac{2}{(4\pi t)^{3/2}} e^{-t} \sum_q \frac{\log q}{q - q^{-1}} e^{-(\log q)^2/4t}$$

where the sum is over $q = q_\gamma$ for $\gamma \neq \mathrm{id}$. Thus the identity term must be handled separately. The integral transformation considered in (6) is sufficiently simple so that we have

$$\frac{2s - 2}{(4\pi)^{3/2}} \int_0^{\infty} e^{-(s-1)^2 t} t^{1/2} \frac{dt}{t} = \frac{1}{4\pi}.$$

As a result, when considering the series (10) for $x = y$, summing over

$$\mathbf{q}_\gamma(x, x) = d_{\mathbf{M}_3}(\tilde{x}, \gamma\tilde{x})$$

with $\gamma \neq \mathrm{id}$, all properties (i) through (iv) hold with property (iii) being changed to allow an additive factor which is a polynomial of degree three in s. The coefficients of this polynomial can be determined by considering the limit as $s \to \infty$ and using the asymptotic formulas from Chapter I.

Remark 1. Computations similar to the above hold for compact quotients of odd dimensional hyperbolic spaces, since, for these manifolds, one can obtain simple expressions similar to (1) for their heat kernels; see page 151 of [Ch 84]. For non-compact quotients, one must take into account the appearance of the Eisenstein series in the spectral decomposition of the heat kernels, and the subsequent appearance of other terms in the additive functional equation. Examples of such formulas, as well as the more complicated situation of even dimensional hyperbolic manifolds, will be treated in a future publication.

Remark 2. Now that we have constructed a zeta function with additive functional equation and a Dirichlet series, we can apply the methods of [JoL 93c] to study the function

$$(11) \qquad \sum_{n=0}^{\infty} \phi_n(x)\phi_n(y)e^{z \cdot i\sqrt{\lambda_n - 1}},$$

defined for $\mathrm{Im}(z) > 0$. As in [JoL 93c], there is a meromorphic continuation of (11) to complex z with singularities at the points $\{\pm \log \mathbf{q}\}$. This should be viewed as a Duistermaat-Guillemin type theorem, as in [DG 75], since (11) is a wave kernel. We shall deal with this situation at greater length elsewhere, especially since it requires a systematic exposition of the additive fundamental class and its corresponding Cramér theorem and explicit formulas.

Remark 3. The above example shows the necessity of an "additive class" of zeta functions, as discussed in the introduction, by which we mean functions which are meromorphic with simple poles, have Dirichlet series in a half-plane, and have an additive functional equation with additive fudge factors expressible as linear combinations of regularized harmonic series types. The above example is not expressible in the fundamental multiplicative class since the residues, as determined in (iv) above, will certainly not be integers.

Remark 4. Readers will note the distinction between the zeta function defined by Minakshisundaram-Pleijel in [MP 49] by the Dirichlet series

$$\sum_{k=0}^{\infty} \frac{\phi_k(x)\phi_k(y)}{\lambda_k^s}$$

and the new zeta function $\Pi'_{x,y}$. The Minakshisundaram-Pleijel zeta function is essentially the one obtained as the Mellin transform of the theta series satisfying the basic theta conditions, whereas the new zeta function is obtained as the Gauss transform of the theta series.

CHAPTER VI

A generalization of Fujii's theorem

Let $L_{\mathbf{Q}}^+$ denote the set of zeros of the Riemann zeta function with positive imaginary part, meaning

$$L_{\mathbf{Q}}^+ = \{\rho \in \mathbf{C} : \zeta_{\mathbf{Q}}(\rho) = 0 \ \text{ and } \ \operatorname{Im}(\rho) > 0\}.$$

Write $\rho = \beta + i\gamma$ for any $\rho \in L_{\mathbf{Q}}^+$. The zeta function

$$(1) \qquad \sum_{\rho \in L_{\mathbf{Q}}^+} \frac{v(\rho)}{\operatorname{Im}(\rho)^s} = \sum_{\rho \in L_{\mathbf{Q}}^+} \frac{v(\rho)}{\gamma^s},$$

defined for $\operatorname{Re}(s) > 1$, was studied in [Del 66] and [Gu 45], and it was shown that (1) admits a meromorphic continuation to all $s \in \mathbf{C}$ with explicitly computable singularities, including a double pole at $s = 1$. Building on these results, Fujii considered the zeta functions

$$(2) \qquad \sum_{\rho \in L_{\mathbf{Q}}^+} v(\rho) \frac{\sin(\alpha\gamma)}{\gamma^s}$$

and

$$(3) \qquad \sum_{\rho \in L_{\mathbf{Q}}^+} v(\rho) \frac{\cos(\alpha\gamma)}{\gamma^s}$$

for non-zero $\alpha \in \mathbf{R}$ and $\operatorname{Re}(s) > 1$. It was shown in [Fu 83] that (2) admits a holomorphic continuation to all $s \in \mathbf{C}$ for any non-zero α, and (3) admits a mermorphic continuation to all $s \in \mathbf{C}$ with a simple pole at $s = 1$ having residue equal to $-(2\pi)^{-1}\Lambda(e^\alpha)e^{-\alpha/2}$, where

$$(4) \quad \Lambda(x) = \begin{cases} \log p, & \text{if } x = p^k \text{ where } p \text{ is a prime and } k \in \mathbf{Z}_{>0} \\ 0, & \text{otherwise.} \end{cases}$$

Instead of functions formed separately with sine and cosine, one may as well consider what we call the **Fujii function**

$$F(s; \alpha) = \sum_{L_{\mathbf{Q}}^+} v(\rho) \frac{e^{i\alpha\gamma}}{\gamma^s}$$

and its meromorphic continuation. We should note that Fujii's proof of the meromorphic continuation involves a very detailed study of many integrals arising from a generalization of Delsarte's work [Del 66] involving various integral transforms of the classical Riemann-von Mangoldt formula.

As remarked on page 233 of [Fu 83], one can prove analogous results for the eigenvalues of the Laplace-Beltrami operator on the fundamental domain of the modular group $PSL(2, \mathbf{Z})$. These theorems are given in [Fu 84b] and are as follows.

Let $\{\lambda_k\}$ be the set of eigenvalues of the hyperbolic Laplacian on the space $PSL(2, \mathbf{Z}) \backslash \mathbf{h}$ and set $\lambda_j = 1/4 + r_j^2$ with $r_j > 0$. For any $\alpha \in \mathbf{R}^+$, Fujii considered the function

(5)
$$\sum_{r_j} v(r_j) \frac{\sin(\alpha r_j)}{r_j^s},$$

which is defined for $\text{Re}(s) > 2$. Through a rather lengthy and involved application of the Selberg trace formula, it was proved in [Fu 84b] that (5) has an analytic continuation to all $s \in \mathbf{C}$ to a holomorphic function. Again, one could consider the Fujii function

$$\sum_{r_j} v(r_j) \frac{e^{i\alpha r_j}}{r_j^s}$$

for all $\alpha \in \mathbf{R}$ and study its meromorphic continuation, thus capturing many of the results obtained by Fujii in the papers [Fu 84b] and [Fu 88a].

At this point, one could envision a series of articles in which one would define and study a Fujii function associated to every special zeta function, such as zeta functions and L-functions from the theory of modular forms, zeta functions and L-functions of number fields, spectral zeta functions constructed from the eigenvalues associated to the Laplacian acting on any non-compact arithmetic Riemann surface, to name a few examples, with each example yielding a new paper. Such a case-by-case extension of the classical Cramér theorem [Cr 19] has begun to appear in the literature. However, in [JoL 93c], we gave a vast generalization of Cramér's theorem containing all previously known special cases and many more.

Similarly, in this chapter, we obtain a generalization of Fujii's theorem which applies to any zeta function with Euler sum and functional equation whose fudge factors are of regularized product type. This generalization is simply a corollary of the generalized Cramér theorem, and, in particular, applies both to the zeta functions arising from algebraic number theory and to those arising from spectral theory.

In §1 we will state the generalized Fujii theorem, and the proof will be given in §2. To conclude this chapter, we will give various examples of the generalized Fujii theorem in §3.

§1. Statement of the generalized Fujii theorem.

Let us assume the notation of the previous chapters. With this, we can state the following result, which we call the generalized **Fujii theorem.**

Theorem 1.1. *Let* $(Z, \tilde{Z}, \Phi)$ *be in the fundamental class. Let* a *be such that* $\sigma_0 + a > \sigma_0'$, *and let* $\{\rho\}$ *be the set of zeros and poles of* Z *in the open infinite rectangle* $\mathcal{R}_a$ *with vertices at the four points*

$$-a + i\infty, \quad -a, \quad \sigma_0 + a, \quad \sigma_0 + a + i\infty.$$

Let $v(\rho) = \mathrm{ord}_\rho Z$ *and set* $\{\lambda\} = \{\rho/i\}$. *Then:*

i) *For any non-zero* $\alpha \in \mathbf{R}$, *the Fujii function*

$$F_{Z,a}(s; \alpha) = \sum_{\rho \in \mathcal{R}_a} v(\rho) \frac{e^{i\alpha\lambda}}{\lambda^s}$$

has a meromorphic continuation to all $s \in \mathbf{C}$.

ii) *For any non-zero* $\alpha \in \mathbf{R}$, *the continuation of the Fujii function* $F_{Z,a}(s; \alpha)$ *is holomorphic for all* $s \in \mathbf{C}$ *except for simple pole at* $s = 1$ *with residue*

$$\begin{cases} -(2\pi)^{-1} c(\mathbf{q}) \log \mathbf{q} & \text{if } \alpha = \log \mathbf{q} \\ -(2\pi)^{-1} c(\tilde{\mathbf{q}})(\log \tilde{\mathbf{q}}) \tilde{\mathbf{q}}^{-\sigma_0} & \text{if } \alpha = -\log \tilde{\mathbf{q}} \\ 0 & \text{otherwise.} \end{cases}$$

In [Fu 83], Fujii considered the zeta function formed with the imaginary parts of the non-trivial zeros of the Riemann zeta function, assuming the Riemann hypothesis. The following theorem generalizes this result.

Theorem 1.2. *With notation as in Theorem 1.1, assume there is a real constant* β_0 *such that* $\rho = \beta_0 + i\gamma$ *for all* $\rho \in \mathcal{R}_a$. *Then:*

i) *For any non-zero* $\alpha \in \mathbf{R}$, *the Fujii function*

$$F_{Z,a}^{RH}(s; \alpha) = \sum_{\rho \in \mathcal{R}_a} v(\rho) \frac{e^{i\alpha\gamma}}{\gamma^s}$$

has a meromorphic continuation to all $s \in \mathbf{C}$.

ii) *For any non-zero $\alpha \in \mathbf{R}$, the continuation of the Fujii function $F_{Z,a}^{RH}(s;\alpha)$ is holomorphic for all $s \in \mathbf{C}$ except for simple pole at $s = 1$ with residue*

$$
\begin{cases}
-(2\pi)^{-1}c(\mathbf{q})(\log \mathbf{q})\mathbf{q}^{-\beta_0} & \text{if } \alpha = \log \mathbf{q} \\
-(2\pi)^{-1}c(\tilde{\mathbf{q}})(\log \tilde{\mathbf{q}})\tilde{\mathbf{q}}^{-\sigma_0+\beta_0} & \text{if } \alpha = -\log \tilde{\mathbf{q}} \\
0 & \text{otherwise.}
\end{cases}
$$

One obtains the Fujii theorem by considering the above series with α and $-\alpha$ since, for example, if $Z = \zeta_{\mathbf{Q}}$, we have

$$
F_{Z,a}(s;\alpha) - F_{Z,a}(s;-\alpha) = 2i \sum_{\rho \in L_{\mathbf{Q}}^+} v(\rho)\frac{\sin(\alpha\gamma)}{\gamma^s}.
$$

Also, as remarked on page 23 of [Fu 84a], we obtain a meromorphic continuation of the series

$$
2 \sum_{\rho \in L_{\mathbf{Q}}^+} v(\gamma)\frac{\cos(\alpha\gamma)}{\gamma^s} = F_{Z,a}(s;\alpha) + F_{Z,a}(s;-\alpha),
$$

both with and without a Riemann hypothesis type assumption.

Thus what appeared up to now to be a phenomenom associated to more or less arithmetic situations, for instance the location of poles at the logs of prime powers or their analogues for the Selberg zeta function, is now seen to be quite a general property of our broad class of functions.

Directly from Theorem 1.2, we have the following corollary.

Corollary 1.3. *In addition to the above conditions, assume $Z = \tilde{Z}$, and assume all coefficients $c(\mathbf{q})$ are real. If all zeros of Z in $\mathcal{R}_a$ lie on a vertical line $\mathrm{Re}(s) = \beta_0$, then we necessarily have $\beta_0 = \sigma_0/2$.*

Finally, let us note that the case of $\alpha = 0$ is handled by our Cramér theorem, specifically Corollary 1.3 of [JoL 93c], and our results from [JoL 93a], specifically Theorem 1.8 and Corollary 1.10. For completeness, let us list this theorem and refer to the above mentioned references in our work for a proof.

Theorem 1.4. *With notation as in Theorem 1.1, assume that* Φ *has reduced order* (M, m). *Then the zeta function*

$$\sum_{\rho \in \mathcal{R}_a} v(\rho) \frac{1}{\lambda^s},$$

which is defined for $\mathrm{Re}(s) > M + 1$, *has a meromorphic continuation to all* $s \in \mathbf{C}$. *If there is a constant* β_0 *such that* $\rho = \beta_0 + i\gamma$ *for all* $\rho \in \mathcal{R}_a$, *then the zeta function*

$$\sum_{\rho \in \mathcal{R}_a} v(\rho) \frac{1}{\gamma^s},$$

which is defined for $\mathrm{Re}(s) > M + 1$, *has a meromorphic continuation to all* $s \in \mathbf{C}$.

The asymptotic expansion in §5 of [JoL 93c] and Corollary 1.10 of [JoL 93a] combine to give an explicit description of the poles of the zeta functions in Theorem 1.4. We will not state these results here, but will simply remark that the location and order of poles of the zeta functions in Theorem 1.4 are determined by the asymptotic expansion near $t = 0$ of the theta function associated to the fudge factor Φ.

§2. Proof of Fujii's theorem.

Let $z = \alpha + it$ for any non-zero $\alpha \in \mathbf{R}$, and, with notation as above, let $\rho = i\lambda$. In [JoL 93c] we proved various analytic properties of the Cramér function

$$V_Z(z) = \sum_{\rho \in \mathcal{R}_a} v(\rho)e^{\rho z},$$

which is defined when $\mathrm{Im}(z) > 0$. In particular, Theorem 1.1 of [JoL 93c] and subsequent discussion imply that the function

$$(1) \qquad 2\pi i V_{Z,a}(z) - e^{\sigma_0 z}\int_{\sigma_0+a-i\infty}^{\sigma_0+a} e^{-sz}\Phi'/\Phi(s)ds$$

has a meromorphic continuation to all $z \in \mathbf{C}$, whose only singularities are simple poles at the points $\log \mathbf{q}$ and $-\log \tilde{\mathbf{q}}$. The residues of these poles are given on page 390 of [JoL 93c].

Now assume that Φ is of regularized product type of reduced order (M, m). By combining Lemma 3.1, Proposition 3.2, Lemma 3.3, and, quite importantly, Lemma 4.2 of [JoL 93c], we conclude that the integral in (1) has a holomorphic continuation to any non-zero $z \in \mathbf{R}$. Therefore, for $\alpha \neq 0$, the function

$$V_{\alpha,Z,a}(t) = V_{Z,a}(\alpha + it) = \sum_{\rho \in \mathcal{R}_a} v(\rho)e^{i\alpha\lambda}e^{-t\lambda}$$

has an asymptotic behavior of the form

$$(2) \qquad V_{\alpha,Z,a}(t) \sim \sum_{n=-1}^{\infty} c_n(\alpha)t^n \qquad \text{as } t \to 0,$$

for some constants $c_n(\alpha)$ which depend on α. Further, from the formula in Theorem 1.1 in [JoL 93c], we have

$$(3) \qquad c_{-1}(\alpha) = \begin{cases} -(2\pi)^{-1}c(\mathbf{q})\log \mathbf{q} & \text{if } \alpha = \log \mathbf{q} \\ -(2\pi)^{-1}c(\tilde{\mathbf{q}})(\log \tilde{\mathbf{q}})\tilde{\mathbf{q}}^{-\sigma_0} & \text{if } \alpha = -\log \tilde{\mathbf{q}} \\ 0 & \text{otherwise.} \end{cases}$$

By applying the Mellin transform, we conclude from (2) that the function

$$\Gamma(s) \sum_{\rho \in \mathcal{R}_a} v(\rho) \frac{e^{i\alpha\lambda}}{\lambda^s} = \int_0^\infty V_{Z,a}(\alpha + it)t^s \frac{dt}{t} = \Gamma(s)\mathbf{M}V_{\alpha,Z,a}(s)$$

has a meromorphic continuation to all $s \in \mathbf{C}$ whose only singularities are simple poles at the points $s \in \mathbf{Z}_{\leq 0}$ and $s = 1$ (see Theorem 1.5 of [JoL 93a]). Since $\Gamma(s)$ has simple poles at these points $s \in \mathbf{Z}_{\leq 0}$, Theorem 1.1 follows.

Let us now assume that any zero or pole ρ of Z in $\mathcal{R}_a$ is such that $\text{Re}(\rho) = \beta_0$, for fixed β_0. Then we can write $\rho = \beta_0 + i\gamma$, so $\lambda = \gamma - i\beta_0$. With $z = \alpha + it$, we have

$$e^{\rho z} = e^{\alpha\beta_0} e^{i\beta_0 t} \cdot e^{i\gamma\alpha} e^{-t\gamma}.$$

Therefore, the function

(4) $$\sum_{\rho \in \mathcal{R}_a} v(\rho) e^{i\gamma\alpha} e^{-\gamma t} = e^{-\alpha\beta_0} e^{-i\beta_0 t} V_{\alpha,Z,a}(t)$$

has asymptotics as in (2). With this, the proof of Theorem 1.2(i) is completed by applying the Mellin transform and the argument given above. Finally, from (3) and (4), one has the asymptotic formula

$$\sum_{\rho \in \mathcal{R}_a} v(\rho) e^{i\gamma\alpha} e^{-\gamma t} \sim e^{-\alpha\beta_0} c_{-1}(\alpha) t^{-1} \quad \text{as } t \to 0,$$

which completes the proof of Theorem 1.2(ii).

The proof of Corollary 1.3 is as follows. Since all numbers γ are real, we have

$$F_{Z,a}^{RH}(s;\alpha) = \sum_{\rho \in \mathcal{R}_a} v(\rho) \frac{\cos(\alpha\gamma)}{\gamma^s} + i \sum_{\rho \in \mathcal{R}_a} v(\rho) \frac{\sin(\alpha\gamma)}{\gamma^s}.$$

If we assume that all numbers $c(\mathbf{q})$ and $\mathbf{q}$ are real, then the residue of the only possible pole of $F_Z^{RH}(s;\alpha)$ is real, hence the series

$$\sum_{\rho \in \mathcal{R}_a} v(\rho) \frac{\sin(\alpha\gamma)}{\gamma^s}$$

has a holomorphic continuation to all $s \in \mathbf{C}$. However, we can also express this series as

$$2i \sum_{\rho \in \mathcal{R}_a} v(\rho) \frac{e^{i\alpha\gamma}}{\gamma^s} = F_{Z,a}^{RH}(s; \alpha) - F_{Z,a}^{RH}(s; -\alpha),$$

which means that the residues at $s = 1$ necessarily cancel, hence $\beta_0 = \sigma_0/2$.

§3. Examples.

As before, specific applications of our general theorems yield several classical theorems, many recent results, and new applications. We list a few examples here.

Example 1: The sine function. Let $Z = \sin(\pi i s)$. In this case, the associated Fujii function is the Dirichlet-like L-function

$$(1) \qquad F_Z(s; \alpha) = \sum_{n=1}^{\infty} \frac{e^{i\alpha n}}{n^s}$$

since $\{q\} = \{e^{2\pi n}\}$ and $c(e^{2\pi n}) = 1/n$. Theorem 1.2 states that the series (1) has a holomorphic continuation whenever $\alpha \neq 2\pi n$ for some $n \in \mathbf{Z}$, and when $\alpha = 2\pi n$, the series (1) has a simple pole at $s = 1$ with residue 1.

Example 2: The Riemann zeta function. Let $Z = \zeta_{\mathbf{Q}}$. If we apply Theorem 1.2 to the Riemann zeta function, assuming the Riemann hypothesis, we obtain the Fujii theorem from [Fu 83]. However, without assuming the Riemann hypothesis, we do have the following result. Let S denote the non-trivial zeros of the Riemann zeta function in the upper half plane, and let $c \in \mathbf{C}$. Then the function

$$\sum_{\rho \in S} e^{(\rho + ic)z} = e^{icz} \sum_{\rho \in S} e^{\rho z}$$

satisfies the asymptotic axiom **AS 2**. Therefore, the proof of Theorem 1.1 implies that for non-zero $\alpha \in \mathbf{R}$, the function

$$(2) \qquad \sum_{\rho \in S} \frac{e^{\alpha \rho / i}}{(\rho/i) + c)^s}$$

has a meromorphic continuation to all $s \in \mathbf{C}$. The only singularity of the continuation of (2) is a simple pole at $s = 1$ and then only when $\alpha = \pm \log p^n$ where p is a prime.

Without any modification, the above argument applies to Dirichlet L-functions and L-functions of number fields. A list of zeta functions for which the Cramér theorem of [JoL 93c] holds is given in section 7 of [JoL 93c].

Example 3: Selberg zeta functions of compact Riemann surfaces. If Z is the Selberg zeta function associated to a compact Riemann surface, then the theta function coming from the Cramér theorem is

$$\sum e^{(1/2 + i\sqrt{\lambda_j - 1/4})z},$$

where λ_j is an eigenvalue of the Laplacian. Hence, this theta function can be viewed as a type of trace of the wave operator.

Example 4: Selberg zeta functions of non-compact Riemann surfaces. As in [Fu 84b], consider the Selberg zeta function associated to $PSL(2, \mathbf{Z})\backslash\mathbf{h}$. It is known that the set of zeros of the Selberg zeta function is the union of two sets: One set being the eigenvalues of the Laplacian, as described above, and the other set associated to the zeros of the Riemann zeta function (see pages 498 and 508 of [He 83]). From the meromorphy of (2), we conclude that the Fujii function formed with the eigenvalues of the Laplacian acting on $PSL(2, \mathbf{Z})\backslash\mathbf{h}$ has a meromorphic continuation to all $\mathbf{C}$. This theorem is the main result of [Fu 84b].

Similarly, since the scattering determinant for any congruence group is expressible in terms of Dirichlet L-series (see [He 83]), the above argument applies to yield the analogue of the Fujii theorem in these cases. The case of a general non-compact hyperbolic Riemann surface, including those associated to non-congruence subgroups, will be considered in [JoL 94].

BIBLIOGRAPHY

[Ba 81] BARNER, K.: On Weil's explicit formula. *J. reine angew. Math.* **323,** 139-152 (1981).

[Ba 90] BARNER, K.: Einführung in die Analytische Zahlentheorie. Preprint (1990).

[BGV 92] BERLINE, N., GETZLER, E., and VERGNE, M.: *Heat Kernels and Dirac Operators*, Grundlehren der mathematicschen Wissenschaften **298** , New York: Springer-Verlag (1992).

[CoG 93] CONREY, J.B., and GHOSH, A.: On the Selberg class of Dirichlet series: small weights. *Duke Math. J.* **72** (1993), 673-693.

[Cr 19] CRAMÉR, H.: Studien über die Nullstellen der Riemannschen Zetafunktion. *Math. Z.* 4 (1919), 104-130.

[DGM 76] DEBIARD, A., GAVEAU, B., and MAZET, E.: Théorèms de Comparison en Géometrie Riemannienne. *Publ. RIMS Kyoto Univ.* **12** (1976) 391-425.

[Del 66] DELSARTE, J.: Formules de Poisson avec reste. *J. Analyse Math.* **17** (1966) 419-431.

[Den 92] DENINGER, C.: Local L-factors of motives and regularized products. *Invent. Math.* **107,** (1992) 135-150.

[Den 93] DENINGER, C.: Lefschetz trace formulas and explicit formulas in analytic number theory. *J. reine angew. Math.* **441** (1993) 1-15.

[DG 75] DUISTERMAAT, J., and GUILLEMIN, V.: The spectrum of positive elliptic operators and periodic bicharacteristics. *Invent. Math.* **29,** (1975) 39-79.

[Fu 83] FUJII, A.: The zeros of the Riemann zeta function and Gibbs's phenomenon. *Comment. Math. Univ. St. Paul* 32 (1983), 99-113.

[Fu 84a] FUJII, A.: Zeros, eigenvalues, and arithmetic. *Proc. Japan Acad. 60 Ser. A.* (1984), 22-25.

[Fu 84b] FUJII, A.: A zeta function connected with the eigenvalues of the Laplace-Beltrami operator on the fundamental domain of the modular group. *Nagoya Math. J.* **96** (1984) 167-174.

[Fu 88a] FUJII, A.: Arithmetic of some zeta function connected with the eigenvalues of the Laplace-Beltrami operator. *Adv. Studies in Pure Math.* **13** (1988) 237-260.

[Fu 88b] FUJII, A.: Some generalizations of Chebyshev's conjecture. *Proc. Japan Acad.* 64 *Ser. A.* (1988) 260-263.

[Fu 93] FUJII, A.: Eigenvalues of the Laplace-Beltrami operator and the von-Mangoldt function. *Proc. Japan Acad.* 69 *Ser. A.* (1993), 125-130.

[Gal 84] GALLAGHER, P. X.: Applications of Guinand's formula. pp 135-157, volume 70 of *Progress in Mathematics* Boston: Birkhauser (1984).

[Gu 45] GUINAND, A. P.: A summation formula in the theory of prime numbers. *Proc. London Math. Soc* (2) **50** (1945) 107-119.

[Hej 76] HEJHAL, D.: *The Selberg Trace Formula for $PSL(2, \mathbf{R})$, volume 1.* Lecture Notes in Mathematics vol. **548** New York: Springer-Verlag (1976).

[Hej 83] HEJHAL, D.:The Selberg trace formula for $PSL(2, \mathbf{R})$, vol. 2. Springer Lecture Notes in Mathematics **1001** (1983).

[In 32] INGHAM, A. E.: *The Distribution of Prime Numbers,* Cambridge University Press, Cambridge, (1932).

[JoL 93a] JORGENSON, J., and LANG, S.: Complex analytic properties of regularized products and series. Springer Lecture Notes in Mathematics **1564** (1993), 1-88.

[JoL 93b] JORGENSON, J., and LANG, S.: A Parseval formula for functions with a singular asymptotic expansion at the origin. Springer Lecture Notes in Mathematics **1564** (1993), 1-88.

[JoL 93c] JORGENSON, J., and LANG, S.: On Cramér's theorem for general Euler products with functional equation. *Math. Ann.* **297** (1993), 383-416.

[JoL 94] JORGENSON, J., and LANG, S.: Applications of explicit formulas to scattering determinants of finite volume hyperbolic Riemann surfaces. In preparation.

[Kur 91] KUROKAWA, N.: Multiple sine functions and Selberg zeta functions. *Proc. Japan Acad., Ser A* **67**, (1991) 61-64.

[La 70] LANG, S.: *Algebraic Number Theory,* Menlo Park, CA.: Addison-Wesley (1970), reprinted as Graduate Texts in Mathematics **110**, New York: Springer-Verlag (1986); third edition, Springer-Verlag (1994).

[La 87] LANG, S.: *Elliptic Functions, second edition.* Graduate Texts in Mathematics **112** New York: Springer-Verlag (1987).

[La 93a] LANG, S.: *Complex Analysis,* Graduate Texts in Mathematics **103**, New York: Springer-Verlag (1985), Third Edition (1993).

[La 93b] LANG, S.: *Real and Functional Analysis, Third Edition,* New York: Springer-Verlag (1993).

[Lav 93] LAVRIK, A. F.: Arithmetic equivalents of functional equations of Riemann type. *Proc. Steklov Inst. Math.* **2** (1993) 237-245.

[MiP 49] MINAKSHISUNDARAM, S., and PLEIJEL, A.: Some properties of the eigenfunctions of the Laplace operator on Riemannian manifolds. *Canadian Jour. Math.* **1** (1949) 242-256.

[Mo 76] MORENO, C. J.: Explicit formulas in automorphic forms. *Lecture Notes in Mathematics* **626** Springer-Verlag (1976).

[Se 56] SELBERG, A.: Harmonic analysis and discontinous groups in weakly symmetric Riemannian spaces with applications to Dirichlet series, *J. Indian Math. Soc. B.* **20** (1956) 47-87 (*Collected papers volume I,* Springer-Verlag (1989) 423-463).

[Sel 91] SELBERG, A.: Old and new conjectures and results about a class of Dirichlet series. *Collected papers volume II,* Springer-Verlag: New York (1991) 47-63.

[St 90] STRICHARTZ, R. S.: Book review of *Heat Kernels and Spectral Theory*, by E. B. Davies, *Bull. Amer. Math. Soc.* **23** (1990) 222-227.

[Ti 48] TITCHMARSH, E. C.: *Introduction to the Theory of Fourier-Integrals, 2nd Edition* Oxford University Press, Oxford (1948).

[Ve 78a] VENKOV, A. B.: A formula for the Chebyshev psi function, *Math. Notes of USSR* **23** (1978) 271-274.

[Ve 78b] VENKOV, A. B.: Selberg's trace formula for the Hecke operator generated by an involution, and the eigenvalues of the Laplace-Beltrami operator on the fundamental domain of the modular group $PSL(2, \mathbf{Z})$. *Math. USSR Izv.* **42** (1978) 448-462.

[Ve 81] VENKOV, A. B.: Remainder term in the Weyl-Selberg asymptotic formula. *J. Soviet Math.* **17** (1981) 2083-2097.

[We 52] WEIL, A.: Sur les "formules explicites" de la théorie des nombres premiers, *Comm. Lund* (vol. dédié à Marcel Riesz), 252-265 (1952).

[We 72] WEIL, A.: Sur les formules explicites de la théorie des nombres, *Izv. Mat. Nauk (Ser. Mat.)* **36**, 3-18 (1972).

A SPECTRAL INTERPRETATION OF WEIL'S EXPLICIT FORMULA

Dorian Goldfeld

§1. Introduction:

The classical theory of automorphic functions over $\mathbb{Q}$ is based on Selberg's spectral decomposition of the space

$$\mathcal{L}^2(GL(2, \mathbb{Q})\backslash GL(2, \mathbb{A}))$$

where $\mathbb{A}$ denotes the adele group of $\mathbb{Q}$. It is well known that this space has an orthogonal decomposition into cusp forms, Eisenstein series, and residues of Eisenstein series. Further, each of these basic functions has a Fourier expansion which defines a canonical zeta function satisfying properties similar to the classical Riemann zeta function. The Rankin-Selberg [R], [S1] convolution of an automorphic form for $GL(2, \mathbb{Q})$ yields a new zeta function associated to an automorphic function on $GL(3, \mathbb{Q})$. This is the Gelbart-Jacquet [G-J] lift. By studying the spectral decomposition of a kernel function on $GL(2, \mathbb{Q})$, Selberg [S2] went considerably further and found the trace formula. This led to the discovery of the Selberg zeta function whose explicit formula is precisely the trace formula. The zeros of the Selberg zeta function determine the discrete spectrum of the Laplacian on the space $\mathcal{L}^2(GL(2, \mathbb{Q})\backslash GL(2, \mathbb{A}))$. The fact that the Laplacian is a self adjoint operator gave the analogue of the Riemann hypothesis for the Selberg zeta function.

The classical theory of automorphic forms is based on the spectral properties of the group $GL(2, \mathbb{Q})$ acting on $GL(2, \mathbb{A})$. We think of this group as generated by additions and one inversion. The object of this paper is to show that an analogue of the classical theory of automorphic forms exists for a group defined over $\mathbb{Q}$ generated by

---

Research supported in part by NSF grant no. DMS 9003907

multiplications and one inversion which acts on $A^\times$, the idele group over $\mathbb{Q}$. The space of $\mathcal{L}^2$ functions on the factor space has an orthogonal decomposition into cusp forms, Eisenstein series, and residues of Eisenstein series. Each of these basic functions has a Mellin expansion which defines a canonical zeta function, and there is a generalization of the Rankin-Selberg method and Gelbart-Jacquet lift. The trace formula for this space is precisely the explicit formula of A. Weil [W], and the discrete spectrum of the Laplacian is given by the zeros of the Riemann zeta function which is just the Selberg zeta function for this space. Our method gives the first interpretation of the integral involving the logarithmic derivative of the gamma function (in Weil's explicit formula) as an integral over the continuous spectrum of a Laplace operator. The Riemann hypothesis remains unproven, however, since it is not known that the analogue of the Petersson inner product for this space is positive definite on the cuspidal spectrum. In fact, the Riemann hypothesis is equivalent to the fact that the space generated by the cusp forms is positive definite. All that we are able to show at present is that this is an indefinite space. If the Riemann zeta function has a zero off the line $\mathrm{Re}(s) = \frac{1}{2}$ then the cusp form associated to this zero will have norm zero.

Manin [M] asks if there exists a category where one can define "absolute Descartes powers,"

$$\mathrm{Spec}\,\mathbb{Z} \times \cdots \times \mathrm{Spec}\,\mathbb{Z}.$$

Following Kurokawa [K], he shows that at least the zeta function of such an object can be defined which agrees with Deninger's [D1-2] representation of zeta functions as regularized infinite determinants.

As in Manin [M], we define a left directed family to be a set Λ of complex numbers λ (which are discrete in $\mathbb{C}$ and where each λ occurs with multiplicity $m_\lambda \in \mathbb{C}$) which satisfies

(1) $\forall r \in \mathbb{R}$, $\quad \mathrm{Card}\left\{\lambda \in \Lambda \mid \mathrm{Re}(\lambda) > r\right\} < \infty$,

(2) $\exists \beta > 0$, $\quad$ s.t. $\quad \displaystyle\sum_{\substack{\lambda \in \Lambda \\ \mathrm{Re}(\lambda) \geq -H}} |m_\lambda| = O(H^\beta)$ $\quad$ as $(H \to \infty)$.

The tensor product $\Lambda_1 \otimes \Lambda_2$ of two left directed families is defined to be

$\Lambda_1 \otimes \Lambda_2$

$$= \left\{ \lambda = \lambda_1 + \lambda_2 \,\middle|\, \lambda_1 \in \Lambda_1,\ \lambda_2 \in \Lambda_2,\ m_\lambda = \sum_{\lambda_1 + \lambda_2 = \lambda} m_{\lambda_1} m_{\lambda_2} \right\}.$$

Let Λ, Λ' be two left directed families and let s be a complex variable. The Kurokawa tensor product $\underset{\text{Kur}}{\otimes}$ of the products associated with the directed families is defined by the formula

$$\prod_{\lambda \in \Lambda}(s - \lambda)^{m_\lambda} \underset{\text{Kur}}{\otimes} \prod_{\lambda' \in \Lambda'}(s - \lambda')^{m_{\lambda'}} = \prod_{\lambda \in \Lambda \otimes \Lambda'}(s - \lambda)^{m_\lambda}.$$

By taking logarithmic derivatives this lifts to what we shall call a Kurokawa sum

$$\sum_{\lambda \in \Lambda}(s - \lambda)^{-m_\lambda} \underset{\text{Kur}}{\oplus} \sum_{\lambda' \in \Lambda'}(s - \lambda')^{-m_{\lambda'}} = \sum_{\lambda \in \Lambda \otimes \Lambda'}(s - \lambda)^{m_\lambda}.$$

The author finds it remarkable that this sum which is expected to be associated to the tensor products of "motives," should be precisely the Rankin-Selberg convolution as exemplified in §8 of this paper.

§2. Notation:

For a rational prime p let $||_p$ denote the p-adic valuation on $\mathbb{Q}$ normalized so that $|p|_p = p^{-1}$. Let $\mathbb{Q}_p$, $\mathbb{Z}_p$, U_p be the p-adic completion of $\mathbb{Q}$ with respect to $||_p$, the p-adic integers, and the p-adic units, respectively. If $x_p \in \mathbb{Q}_p$, we set $e(x_p) = \frac{\log(|x_p|_p)}{\log p}$, the exponent of x_p and $s(x_p) = \frac{e(x_p)}{|e(x_p)|}$, the sign of x_p.

Let $\mathbb{A}$ denote the group of adeles over $\mathbb{Q}$, and let $\mathbb{A}^\times$ denote the group of ideles over $\mathbb{Q}$ where each $x \in \mathbb{A}^\times$ is of the form $x = (x_\infty, x_2, x_3, \ldots, x_p, \ldots)$ with $x_\infty \in \mathbb{R}$, $x_p \in \mathbb{Q}_p$, and $x_p \in U_p$ for almost all primes p. For a subset $B \subset \mathbb{Q}_p$ define

$$|x_p|_B = \begin{cases} |x|_p & \text{for } x_p \in B \\ 0 & \text{for } x_p \notin B. \end{cases}$$

We also set $||x|| = \prod_v |x_v|_v$ to be the norm on $\mathbb{Q}^\times$ where $|x|_v = |x|$, the ordinary absolute value when $v = \infty$.

If we consider $\mathbb{Q}^\times$ embedded diagonally in $\mathbb{A}^\times$ then $\mathbb{Q}^\times$ acts on $\mathbb{A}^\times$ by multiplication. For $\alpha \in \mathbb{Q}^\times$, $x \in \mathbb{A}^\times$ define this action by

$$\alpha x = (\alpha x_\infty, \alpha x_2, \alpha x_3, \ldots).$$

Let ω denote the involution $\omega x = x^{-1}$, and define the group

$$\mathfrak{G} =< \mathbb{Q}^{\times}, \omega >$$

as the group generated by $\mathbb{Q}^{\times}$ and ω. Then $\mathfrak{G}$ acts on $\mathbb{A}^{\times}$ and may be realized as a matrix group generated by

$$\left\{ \begin{pmatrix} \alpha & 0 \\ 0 & 1 \end{pmatrix} \, \bigg| \, \alpha \in \mathbb{Q} \right\}$$

and $\begin{pmatrix} 0 & 1 \\ 1 & 0 \end{pmatrix}$. Let Z denote the center of $\mathfrak{G}$, and define the discrete group $\Gamma = \mathfrak{G}/Z$. Then every element of Γ is of the form $\begin{pmatrix} \alpha & 0 \\ 0 & 1 \end{pmatrix}$ or $\begin{pmatrix} 0 & \alpha \\ 1 & 0 \end{pmatrix}$ for $\alpha \in \mathbb{Q}^{\times}$.

§3. Construction of the indefinite space $\mathcal{L}^2(\Upsilon)$:

A function $f : \mathbb{R} \to \mathbb{C}$ is said to be a symmetric Schwartz function if it is smooth and satisfies the conditions

$$f(x) = f(1/x), \quad f(x) = f(-x),$$

$$\left| x^m \frac{d}{dx^m} f(x) \right| = O(1) \qquad \text{(for all } m, n \geq 0, \ n \in \mathbb{Z}\text{)}.$$

Let $\mathcal{S}$ denote the space of symmetric Schwartz functions. For $f \in \mathcal{S}$ the Mellin transform of f is given by

$$\tilde{f}(\lambda) = \int_0^{\infty} f(x) x^{\lambda} \, \frac{dx}{x},$$

so that $f(x)$ is given by the inverse Mellin transform

$$f(x) = \frac{1}{2\pi i} \int_{\sigma - i\infty}^{\sigma + i\infty} \tilde{f}(\lambda) x^{-\lambda} \frac{dx}{x},$$

for suitable σ.

We now construct a vector space Υ over $\mathbb{C}$ with an indefinite inner product. Every element of Υ will be a function $F : \Gamma\backslash\mathbb{A}^\times \to \mathbb{C}$ of the form

$$F(x) = \sum_p (\log p) \sum_{\alpha\in\Gamma} |\alpha x_p|_{\mathbb{Q}_p}^{-\frac{1}{2}s(\alpha x_p)} \cdot \left(\prod_{q\neq p,\infty} |\alpha x_q|_{U_q} \right) \cdot f(\alpha x_\infty),$$

where $f : \mathbb{R} \to \mathbb{C}$ is a symmetric Schwartz function. A simple computation shows that

$$F(x) = \sum_p (\log p) \sum_{n=1}^\infty \frac{1}{p^{n/2}} \left[f(p^n \|x\|) + f(p^{-n}\|x\|) \right].$$

We shall say that F is associated to f. Given $F, G \in \Upsilon$ associated to f, g, respectively, we define an inner product

$$<F,G> = \int_{\Gamma\backslash\mathbb{A}^\times} F(x)\overline{g(\|x\|)} \frac{dx}{\|x\|}.$$

After some computations we obtain

$$\begin{aligned}<F,G> &= \int_{\Gamma\backslash\mathbb{A}^\times} F(x)\overline{g(\|x\|)} \frac{dx}{\|x\|} \\ &= \frac{1}{4\pi i} \int_{2-i\infty}^{2+i\infty} -\frac{\zeta'}{\zeta}\left(\frac{1}{2}+\lambda\right) \tilde{f}(\lambda)\overline{\tilde{g}(\lambda)}\, d\lambda.\end{aligned}$$

It immediately follows that $<F,G> = \overline{<G,F>}$, and thus the inner product endows Υ with the structure of an indefinite inner product space.

§4. Spectral theory of $\mathcal{L}^2(\Upsilon)$:

Let v be a place of $\mathbb{Q}$ and let $\ell \in \mathbb{Q}_v$. Consider the multiplication operator $M_{\ell,v} = M_\ell$, where $M_\ell : \Upsilon \to \Upsilon$ is given by

$$M_\ell F(x) = F((x_\infty, x_2,\dots,\ell x_v,\dots)) + F((x_\infty, x_2,\dots,\frac{1}{\ell}x_v,\dots)).$$

Clearly $< M_\ell F, G > = < F, M_\ell G >$ so the collection of M_ℓ form a commuting system of symmetric operators. Each of these operators commute with the Laplacian

$$\Delta = \left(||x|| \frac{d}{d||x||} \right)^2.$$

We shall show that each operator M_ℓ is a bounded operator on $\mathcal{L}^2(\Upsilon)$ with discrete spectrum

$$|\ell|_v^\rho + |\ell|_v^{-\rho},$$

where $\frac{1}{2} + \rho$ is a zero or pole of the Riemann zeta function. Further, M_ℓ has a continuous spectrum

$$|\ell|_v^\lambda + |\ell|_v^{-\lambda},$$

with $\lambda \in i\mathbb{R}$. Similarly, the Laplacian Δ is an unbounded operator on $\mathcal{L}^2(\Upsilon)$ with discrete spectrum

$$\rho^2 + \rho^{-2},$$

where $\frac{1}{2} + \rho$ is a zero or pole of the Riemann zeta function. The Laplacian Δ also has a continuous spectrum

$$\lambda^2 + \lambda^{-2},$$

with $\lambda \in i\mathbb{R}$. In analogy with the Selberg theory for $SL(2, \mathbb{R})$ the space $\mathcal{L}^2(\Upsilon)$ decomposes into cusp forms, Eisenstein series, and residues of Eisenstein series.

§5. Eisenstein series:

Formally, we define the Eisenstein series as the element of Υ associated to the symmetric function $||x||^s + ||x||^{-s}$, which is given in the form

$$E(x, s) = \sum_p (\log p) \sum_{n=1}^\infty \frac{1}{p^{n/2}} \left[(p^n ||x||)^s + (p^{-n} ||x||)^s \right.$$

$$\left. + (p^n ||x||)^{-s} + (p^{-n} ||x||)^{-s} \right]$$

$$= \left[-\frac{\zeta'}{\zeta} \left(\frac{1}{2} + s \right) - \frac{\zeta'}{\zeta} \left(\frac{1}{2} - s \right) \right] \cdot \left[||x||^s + ||x||^{-s} \right].$$

The only problem is that $||x||^s + ||x||^{-s}$ is not Schwartz, and hence, the above series does not formally converge. The problem can be circumvented by defining the Eisenstein series as a distribution.

Set

$$\omega(s) = -\frac{\zeta'}{\zeta}\left(\frac{1}{2}+s\right) - \frac{\zeta'}{\zeta}\left(\frac{1}{2}-s\right)$$

$$= -\log \pi + \frac{1}{2}\frac{\Gamma'}{\Gamma}\left(\frac{\frac{1}{2}+s}{2}\right) + \frac{1}{2}\frac{\Gamma'}{\Gamma}\left(\frac{\frac{1}{2}-s}{2}\right).$$

We have the identity

$$\omega(s)(x^s+x^{-s}) = \frac{1}{4\pi i}\int_{\sigma-i\infty}^{\sigma+i\infty} \omega(\lambda)\cdot\left(\frac{1}{\lambda+s} + \frac{1}{\lambda-s}\right)(x^\lambda+x^{-\lambda})\,d\lambda,$$

which is valid for $0 < \sigma < \frac{1}{2}$, $\sigma > |Re(s)|$.

Let S denote the test function space on $\Gamma\backslash\mathbf{A}^\times$, and define K to be the space of all smooth functions

$$\Gamma\backslash\mathbf{A}^\times \to \mathbf{C}.$$

Then for $g \in S$, $G \in \mathcal{L}^2(\Upsilon)$, and $F \in K$, we consider the continuous linear functional

$$< F,G > = \int_{\Gamma\backslash\mathbf{A}^\times} F(x)\overline{g(||x||)}\,\frac{dx}{||x||},$$

which defines F as a distribution.

In this manner for $g \in S$ and $G \in \Upsilon$ associated to g define

$$< E,G > = \int_{\Gamma\backslash\mathbf{A}^\times} E(x,s)\overline{g(||x||)}\,\frac{dx}{||x||}$$

$$= \frac{1}{4\pi i}\int_{\sigma-i\infty}^{\sigma+i\infty}\left(-\frac{\zeta'}{\zeta}(\frac{1}{2}+\lambda) - \frac{\zeta'}{\zeta}(\frac{1}{2}-\lambda)\right)\cdot$$

$$\left(\frac{1}{s+\lambda} + \frac{1}{-s+\lambda}\right)\tilde{\bar{g}}(\lambda)\,d\lambda,$$

which is valid for $0 < \sigma < \frac{1}{2}$, $\sigma > |Re(s)|$. This defines $E(x,s)$ as a distribution.

Now, for $f \in S$, define

$$E_F(x) = \frac{1}{4\pi i} \int_{-i\infty}^{+i\infty} E(x,s)\tilde{f}(s)\,ds,$$

the projection of F onto the space of Eisenstein series. Then

$$E_F(x) = O(1),$$

so that E_F is in $\mathcal{L}^2$. We compute

$$< E_F, G >$$

$$= \frac{1}{4\pi i} \int_{-i\infty}^{+i\infty} \tilde{f}(s) \int_{\sigma-i\infty}^{\sigma+i\infty} \omega(\lambda) \cdot \left(\frac{1}{s+\lambda} + \frac{1}{-s+\lambda} \right) \tilde{g}(\lambda)\,d\lambda\,ds$$

$$= \frac{1}{2\pi i} \int_{\sigma-i\infty}^{\sigma+i\infty} \omega(\lambda)\tilde{f}(\lambda)\,\tilde{g}(\lambda)\,d\lambda,$$

which is valid for $0 < \sigma < \frac{1}{2}$. It follows that

$$< E_F, G > = < F, E_G > = \overline{< E_G, F >}.$$

In the same manner, for $0 < \sigma < \frac{1}{2}$, $\sigma > |Re(s)|$,

$$< E_F, E > = \frac{1}{2\pi i} \int_{\sigma-i\infty}^{\sigma+i\infty} \omega(\lambda)\tilde{f}(\lambda) \left(\frac{1}{s+\lambda} + \frac{1}{-s+\lambda} \right) d\lambda$$

$$= \tilde{f}(s)\omega(s)$$

$$= \overline{< E, E_F >}.$$

It follows that

$$< (F - E_F), E > = 0,$$
$$< E, (F - E_F) > = 0,$$

and finally

$$< (G - E_G), (F - E_F) > \; = < G, (F - E_F) >$$

$$= \overline{< (G - E_G), F >}$$

$$= \overline{< (F - E_F), (G - E_G) >}.$$

The above computations establish the fact that $F - E_F$ lies in the orthogonal complement of the space of Eisenstein series.

To recapitulate, the Eisenstein series

$$E(x, s)$$

$$= \left[-\log \pi + \frac{1}{2} \frac{\Gamma'}{\Gamma} \left(\frac{\frac{1}{2} + s}{2} \right) + \frac{1}{2} \frac{\Gamma'}{\Gamma} \left(\frac{\frac{1}{2} - s}{2} \right) \right] \left(||x||^s + ||x||^{-s} \right),$$

may be defined as a distribution on $\mathcal{L}^2(\Upsilon)$ which satisfies

$$\Delta E(x, s) = s^2 E(x, s),$$

$$M_\ell E(x, s) = \left(||\ell||^s + ||\ell||^{-s} \right) E(x, s),$$

and, hence, determines the continuous spectrum of these operators. The Eisenstein series has poles at $s = \frac{1}{2}, -\frac{1}{2}, \frac{3}{2}, -\frac{3}{2}, \frac{5}{2}, -\frac{5}{2}, \ldots$ with residues given by a constant multiple of

$$||x||^\kappa + ||x||^{-\kappa},$$

for $\kappa = \frac{1}{2}, -\frac{1}{2}, \frac{3}{2}, -\frac{3}{2}, \frac{5}{2}, -\frac{5}{2}, \ldots$

§6. Cusp Forms:

Let $\xi(s) = \pi^{-\frac{s}{2}} \Gamma \left(\frac{s}{2} \right) \zeta(s)$. The functional equation of the zeta function may be written in the form

$$\frac{\xi'}{\xi}(s) = -\frac{\xi'}{\xi}(1 - s).$$

Let f be a symmetric Schwartz function. A consequence of the functional equation of the zeta function is the well known explicit formula

$$\sum m_\rho \left(x^\rho + x^{-\rho} \right) \tilde{f}(\rho) =$$

$$\frac{1}{2\pi i} \int_{c-i\infty}^{c+i\infty} -\frac{\xi'}{\xi} \left(\frac{1}{2} + \lambda \right) \left(x^\lambda + x^{-\lambda} \right) \tilde{f}(\lambda) \, d\lambda$$

$$+ \frac{1}{2\pi i} \int_{c-i\infty}^{c+i\infty} -\left[\frac{\xi'}{\xi} \left(\frac{1}{2} + \lambda \right) + \frac{\xi'}{\xi} \left(\frac{1}{2} - \lambda \right) \right] x^\lambda \, \tilde{f}(\lambda) \, d\lambda,$$

where $c > \frac{1}{2}$. The sum on the left in the above formula runs over complex numbers ρ which satisfy

$$\xi(\frac{1}{2} + \rho) = 0, \ \infty.$$

Each zero or pole has multiplicity m_ρ (taken negatively if $\xi(\frac{1}{2}+\rho) = 0$). The only pole is a simple pole at $\rho = \frac{1}{2}$ so that $m_{\frac{1}{2}} = +1$. The only difference between this formula and Weil's [W] explicit formula is that it involves the Mellin transform instead of the Fourier transform. Of course one can easily pass from one to the other by a logarithmic transformation.

Let $F \in \Upsilon$ be associated to f. Replacing x by $||x||$, the norm of an idele (in the above formula), we may rewrite the explicit formula in the form

$$F(x) = \sum_{\xi(\frac{1}{2}+\rho)=0,\infty} m_\rho \tilde{f}(\rho) \left(||x||^\rho + ||x||^{-\rho} \right) - E_F(x).$$

It now follows from the results of §5 that the cusp forms are given by the functions

$$||x||^\rho + ||x||^{-\rho},$$

where $\frac{1}{2} + \rho$ is a zero of the zeta function. In fact these functions can be constructed explicitly by forming $F(x) - E_F(x)$, where F is associated to f and $\tilde{f}(\lambda)$ vanishes at all the zeros or poles of the zeta function with one exception. Every $F \in \mathcal{L}^2(\Upsilon)$ can be expressed as a linear combination of cusp forms plus an integral of Eisenstein series plus a multiple of

$$||x||^{\frac{1}{2}} + ||x||^{-\frac{1}{2}},$$

which is a residue of the Eisenstein series. This establishes the spectral decomposition of the space $\mathcal{L}^2(\Upsilon)$.

It is not hard to show that a cusp form $||x||^\rho + ||x||^{-\rho}$ has norm zero if and only if ρ is not pure imaginary. This establishes the fact that the Riemann hypothesis is equivalent to the positivity of the inner product $< >$ on the cuspidal spectrum.

§7. The zeta function associated to an automorphic form on $\mathcal{L}^2(\Upsilon)$:

Let $F(x) \in \mathcal{L}^2(\Upsilon)$ be associated to $f \in \mathcal{S}$. Then

$$F(x) = \frac{1}{2\pi i} \int_{\sigma-i\infty}^{\sigma+i\infty} -\frac{\zeta'}{\zeta}\left(\frac{1}{2}+\lambda\right) \tilde{f}(\lambda)(x^\lambda + x^{-\lambda})\, d\lambda,$$

for $\sigma > \frac{1}{2}$. We define the zeta function associated to F by the formula

$$\zeta_F(s) = \frac{1}{2\pi i} \int_{\sigma-i\infty}^{\sigma+i\infty} -\frac{\zeta'}{\zeta}\left(\frac{1}{2}+\lambda\right) \tilde{f}(\lambda)\left(\frac{1}{-\lambda+s} + \frac{1}{\lambda+s}\right) d\lambda,$$

which is the Mellin transform $\tilde{F}(s)$. To see this, we note that the Mellin transform of $x^\lambda + x^{-\lambda}$ (denoted by $(\widetilde{x^\lambda+x^{-\lambda}})(s)$) is defined as a distribution by the formula

$$\frac{1}{2\pi i} \int_{\sigma-i\infty}^{\sigma+i\infty} (\widetilde{x^\lambda+x^{-\lambda}})(s)\,\tilde{g}(s)\, ds = \int_0^\infty (x^\lambda + x^{-\lambda})g(x)\,\frac{dx}{x}$$

$$= \frac{1}{2\pi i} \int_{\sigma-i\infty}^{\sigma+i\infty} \left(\frac{1}{s-\lambda} + \frac{1}{s+\lambda}\right)\tilde{g}(s)\, ds$$

$$= 2\tilde{g}(\lambda),$$

which defines a continuous linear functional on $[0, \infty]$ with invariant multiplicative measure $\frac{dx}{x}$. We may thus interpret $-\frac{\zeta'}{\zeta}(\frac{1}{2}+\lambda)\tilde{f}(\lambda)$ as the λth Mellin coefficient of $F(x)$. This is analogous to the Fourier coefficient in the classical theory of automorphic forms where the expansion is taken with respect to the additive group. The zeta function associated to F satisfies the functional equation

$$\zeta_F(s) = -\zeta_F(-s).$$

The zeta function associated to a cusp form $||x||^\rho + ||x||^{-\rho}$ with ρ such that $\zeta(\frac{1}{2} + \rho) = 0$, will be

$$\frac{1}{s-\rho} + \frac{1}{s+\rho}.$$

§8. The Rankin-Selberg convolution:

Let
$$\eta_\rho = ||x||^\rho + ||x||^{-\rho}, \quad \eta_{\rho'} = ||x||^{\rho'} + ||x||^{-\rho'},$$

be two cusp forms where $\zeta(\frac{1}{2} + \rho) = \zeta(\frac{1}{2} + \rho') = 0$. Let $F \in \mathcal{L}^2(\Upsilon)$ be associated to $f \in \mathcal{S}$. We form the inner product

$$< F, \overline{\eta_\rho \eta_{\rho'}} >$$
$$= \int_0^\infty f(||x||) \cdot (||x||^\rho + ||x||^{-\rho}) \cdot (||x||^{\rho'} + ||x||^{-\rho'}) \frac{dx}{||x||}$$
$$= 2\tilde{f}(\rho + \rho') + 2\tilde{f}(\rho - \rho').$$

The above formula can be further generalized by enlarging the allowable class of functions F to include distributions. If we consider the case
$$F(x) = \frac{E(x, s)}{\omega(s)},$$

and recall the previously proved fact that the Mellin transform

$$(\widetilde{x^\lambda + x^{-\lambda}})(s) = \frac{1}{s - \lambda} + \frac{1}{s + \lambda},$$

we obtain the Rankin-Selberg convolution

$$< E(, s), \overline{\eta_\rho \eta_{\rho'}} >$$
$$= 2 \left[\frac{1}{s + \rho + \rho'} + \frac{1}{s - \rho - \rho'} + \frac{1}{s + \rho - \rho'} + \frac{1}{s - \rho + \rho'} \right].$$

This may be interpreted as the zeta function associated to an automorphic function for a certain group acting on $(\mathbf{A}^\times)^2$ in analogy with the Gelbart-Jacquet (see [G-J]) lift. These ideas will be briefly explained in the next section.

§9. Higher rank generalizations:

For $n \geq 2$ let W_n denote the Weyl group of $GL(n)$ (which consists of all $n \times n$ matrices with zero entries except for exactly one

1 in each row and column). For an arbitrary multiplicative group K let $\mathcal{D}_n(K)$ denote the diagonal group of all matrices of the form

$$\begin{pmatrix} k_n & & & \\ & \ddots & & \\ & & k_2 & \\ & & & k_1 \end{pmatrix} \qquad \text{with} \quad k_i \in k \ (1 \leq i \leq n).$$

The Weyl group W_n acts on $\mathcal{D}_n(K)/K$ by conjugation. For a normal subgroup $H \trianglelefteq K$, the group $\mathcal{D}_n(H)/H$ acts on $\mathcal{D}_n(K)/K$ by left matrix multiplication. We define the group $\mathcal{G}_n(H)$ to be the group generated by the group $\mathcal{D}_n(H)/H$ and the Weyl group W_n; by construction $\mathcal{G}_n(H)$ acts naturally on $\mathcal{D}_n(K)/K$.

We now consider the group $\Gamma_n = \Gamma_n(\mathbb{Q}^\times) = \mathcal{G}_n(\mathbb{Q}^\times)/\mathbb{Q}^\times$ which acts on $\mathcal{D}_n(\mathbb{A}^\times)/\mathbb{A}^\times$. Every $X \in \mathcal{D}_n(\mathbb{A}^\times)/\mathbb{A}^\times$ can be put in the canonical form

$$X = \begin{pmatrix} x^{n-1} & & & \\ & \ddots & & \\ & & x^1 & \\ & & & 1 \end{pmatrix},$$

which we write succinctly as $X = (x^1, \dots, x^{n-1})$. Let $X_\infty = (x_\infty^1, \dots, x_\infty^{n-1})$ and $X_p = (x_p^1, \dots, x_p^{n-1})$ denote the infinite and p-component of X, respectively. A function

$$f : W_n \backslash \mathcal{D}_n(\mathbb{R}^\times)/\mathbb{R}^\times \to \mathbb{C},$$

is said to be a Schwartz function if it and its partial derivatives (of all orders) with respect to $x^1, \dots, x^{n-1}$ are bounded. Denote by $\mathcal{S}_n$ the space of all such Schwartz functions.

We shall now construct a vector space (over $\mathbb{C}$) of automorphic functions for the group Γ_n. Let

$$X = (x^1, \dots, x^{n-1}) \in \mathcal{D}_n(\mathbb{A}^\times)/\mathbb{A}^\times.$$

For a prime q, the function

$$\mathfrak{U}_q(X_q) = \left| x_q^1 \right|_{U_q} \cdots \left| x_q^{n-1} \right|_{U_q}$$

is invariant under the group W_n. For an n-tuple of primes

$$P = (p_1, p_2, \ldots, p_n),$$

we adopt the simplifying notation:

$$X_P = \{X_{p_1}, \ldots, X_{p_{n-1}}\},$$
$$\log P = (\log p_1) \cdots (\log p_{n-1}).$$

The function

$$\Delta_P(X_P) = \sum_{w \in W_n} \prod_{i=1}^{n-1} \left| w x_{p_i}^i \right|_{\mathbb{Q}_{p_i} - U_{p_i}}^{-\frac{1}{2} s(x_{p_i}^j)}$$

is automatically invariant under W_n. We define the vector space Υ_n over $\mathbb{C}$ to be all linear combinations (with complex coefficients) of automorphic functions of the form

$$F(X) =$$

$$\sum_{\alpha \in \Gamma_n} \sum_{P=(p_1,\ldots,p_{n-1})} (\log P) \cdot \Delta_P(\alpha X_P) \cdot \left(\prod_{q \notin P} \mathfrak{U}_q(\alpha X_q) \right) \cdot f(\alpha X_\infty)$$

with $f \in \mathcal{S}_n$. We say F is associated to f. As in the rank one case, an inner product on this space may be defined as follows. Let $F, G \in \Upsilon_n$ be associated to $f, g \in \mathcal{S}_n$, respectively. Define the indefinite inner product

$$< F, G > = \int_{\Gamma_n \backslash \mathcal{D}_n(\mathbb{A}^\times)/\mathbb{A}^\times} F(x) \cdot \overline{g(||x^1||, \ldots, ||x^{n-1}||)} \prod_{j=1}^{n-1} \frac{dx^j}{||x^j||}.$$

In the special case where F is associated to an f given in the form

$$\sum_{w \in W_n} ||wx^1||^{s_1} \cdots ||wx^{n-1}||^{s_{n-1}}$$

we obtain a minimal parabolic Eisenstein series which lies in the continuous spectrum of the space of differential operators which commute with our group $\mathfrak{G}_n$.

As an example, we consider the case $n = 3$. The group Γ_3, which acts on $\mathcal{D}_n(\mathbf{A}^\times)/\mathbf{A}^\times$, is generated by the transformations

$$\begin{pmatrix} x^2 & & \\ & x^1 & \\ & & 1 \end{pmatrix} \longrightarrow \begin{pmatrix} \alpha_2 x^2 & & \\ & \alpha_1 x^1 & \\ & & 1 \end{pmatrix} \qquad \text{for} \quad \alpha_1, \alpha \in \mathbb{Q},$$

and the symmetries

$$\begin{pmatrix} x^2 & & \\ & x^1 & \\ & & 1 \end{pmatrix} \mapsto \begin{pmatrix} x^1 & & \\ & x^2 & \\ & & 1 \end{pmatrix}$$

$$\begin{pmatrix} x^2 & & \\ & x^1 & \\ & & 1 \end{pmatrix} \mapsto \begin{pmatrix} \frac{x^2}{x^1} & & \\ & \frac{1}{x^1} & \\ & & 1 \end{pmatrix}$$

$$\begin{pmatrix} x^2 & & \\ & x^1 & \\ & & 1 \end{pmatrix} \mapsto \begin{pmatrix} \frac{x^1}{x^2} & & \\ & \frac{1}{x^2} & \\ & & 1 \end{pmatrix}$$

$$\begin{pmatrix} x^2 & & \\ & x^1 & \\ & & 1 \end{pmatrix} \mapsto \begin{pmatrix} \frac{1}{x^1} & & \\ & \frac{x^2}{x^1} & \\ & & 1 \end{pmatrix}$$

$$\begin{pmatrix} x^2 & & \\ & x^1 & \\ & & 1 \end{pmatrix} \mapsto \begin{pmatrix} \frac{1}{x^2} & & \\ & \frac{x^1}{x^2} & \\ & & 1 \end{pmatrix},$$

associated to the Weyl group elements

$$\begin{pmatrix} 0 & 1 & 0 \\ 1 & 0 & 0 \\ 0 & 0 & 1 \end{pmatrix}, \quad \begin{pmatrix} 1 & 0 & 0 \\ 0 & 0 & 1 \\ 0 & 1 & 0 \end{pmatrix}, \quad \begin{pmatrix} 0 & 0 & 1 \\ 1 & 0 & 0 \\ 0 & 1 & 0 \end{pmatrix},$$

$$\begin{pmatrix} 0 & 1 & 0 \\ 0 & 0 & 1 \\ 1 & 0 & 0 \end{pmatrix}, \quad \begin{pmatrix} 0 & 0 & 1 \\ 0 & 1 & 0 \\ 1 & 0 & 0 \end{pmatrix},$$

respectively. The cusp forms will be of type

$$||x^1||^\rho \cdot ||x^2||^{\rho'} + ||x^1||^{\rho'} \cdot ||x^2||^\rho +$$
$$||x^1||^{-\rho-\rho'} \cdot ||x^2||^{\rho'} + ||x^1||^{\rho'} \cdot ||x^2||^{-\rho-\rho'} +$$
$$||x^1||^{-\rho-\rho'} \cdot ||x^2||^\rho + ||x^1||^\rho \cdot ||x^2||^{-\rho-\rho'},$$

where $\frac{1}{2} + \rho$, and $\frac{1}{2} + \rho'$ are nontrivial zeros of the Riemann zeta function. In this manner, the lift from $\mathcal{G}_2$ to $\mathcal{G}_3$ can be established.

§10. References:

[D1] Ch. Deninger, *On the Γ-factors attached to motives*, Inv. Math. **104** (1991), 245-261.

[D2] Ch. Deninger, *Local L-factors of motives and regularized determinants*, Inv. Math. **107** (1992), 135-150.

[G-J] Gelbart, S. and Jacquet, H. *A relation between automorphic representations of GL(2) and GL(3)*, Ann. Sci. Ecole Normale Sup. 4e série. **11** (1978), 471-552.

[K] Kurokawa, N. *Multiple zeta functions: an example,* in Zeta Functions in Geometry, ed. Kurokawa, N. and Sunada, T., Advanced Studies in Pure Mathematics **21** (1992), 219-226.

[M] Manin, Y. *Lectures on zeta functions and motives*, to appear in Astérisque.

[R] Rankin, R. *Contributions to the theory of Ramanujan's function $\tau(n)$ and similar arithmetic functions, I and II*, Proc. Cambridge Phil. Soc. **35** (1939), 351-356, 357-372.

[S1] Selberg, A. *Bemerkungen über eine Dirichletsche Reihe, die mit der Theorie der Modulformen nahe verbunden ist*, Arch. Math. Naturvid. **43** (1940), 47-50.

[S2] Selberg, A. *Harmonic analysis and discontinuous groups in weakly symmetric Riemannian spaces with applications to Dirichlet series*, J. Indian Math. Soc. **20** (1956), 47-87.

[W] Weil, A. *Sur les 'formules explicites' de la théorie des nombres premiers*, Comm. Sém. Math. Univ. Lund 1952, Tome supplémentaire (1952), 252-265.

Department of Mathematics
Columbia University
New York, NY 10027
goldfeld@columbia.edu

INDEX

Vol. 1551: L. Arkeryd, P. L. Lions, P.A. Markowich, S.R. S. Varadhan. Nonequilibrium Problems in Many-Particle Systems. Montecatini, 1992. Editors: C. Cercignani, M. Pulvirenti. VII, 158 pages 1993.

Vol. 1552: J. Hilgert, K.-H. Neeb, Lie Semigroups and their Applications. XII, 315 pages. 1993.

Vol. 1553: J.-L- Colliot-Thélène, J. Kato, P. Vojta. Arithmetic Algebraic Geometry. Trento, 1991. Editor: E. Ballico. VII, 223 pages. 1993.

Vol. 1554: A. K. Lenstra, H. W. Lenstra, Jr. (Eds.), The Development of the Number Field Sieve. VIII, 131 pages. 1993.

Vol. 1555: O. Liess, Conical Refraction and Higher Microlocalization. X, 389 pages. 1993.

Vol. 1556: S. B. Kuksin, Nearly Integrable Infinite-Dimensional Hamiltonian Systems. XXVII, 101 pages. 1993.

Vol. 1557: J. Azéma, P. A. Meyer, M. Yor (Eds.), Séminaire de Probabilités XXVII. VI, 327 pages. 1993.

Vol. 1558: T. J. Bridges, J. E. Furter, Singularity Theory and Equivariant Symplectic Maps. VI, 226 pages. 1993.

Vol. 1559: V. G. Sprindžuk, Classical Diophantine Equations. XII, 228 pages. 1993.

Vol. 1560: T. Bartsch, Topological Methods for Variational Problems with Symmetries. X, 152 pages. 1993.

Vol. 1561: I. S. Molchanov, Limit Theorems for Unions of Random Closed Sets. X, 157 pages. 1993.

Vol. 1562: G. Harder, Eisensteinkohomologie und die Konstruktion gemischter Motive. XX, 184 pages. 1993.

Vol. 1563: E. Fabes, M. Fukushima, L. Gross, C. Kenig, M. Röckner, D. W. Stroock, Dirichlet Forms. Varenna, 1992. Editors: G. Dell'Antonio, U. Mosco. VII, 245 pages. 1993.

Vol. 1564: J. Jorgenson, S. Lang, Basic Analysis of Regularized Series and Products. IX, 122 pages. 1993.

Vol. 1565: L. Boutet de Monvel, C. De Concini, C. Procesi, P. Schapira, M. Vergne. D-modules, Representation Theory, and Quantum Groups. Venezia, 1992. Editors: G. Zampieri, A. D'Agnolo. VII, 217 pages. 1993.

Vol. 1566: B. Edixhoven, J.-H. Evertse (Eds.), Diophantine Approximation and Abelian Varieties. XIII, 127 pages. 1993.

Vol. 1567: R. L. Dobrushin, S. Kusuoka, Statistical Mechanics and Fractals. VII, 98 pages. 1993.

Vol. 1568: F. Weisz, Martingale Hardy Spaces and their Application in Fourier Analysis. VIII, 217 pages. 1994.

Vol. 1569: V. Totik, Weighted Approximation with Varying Weight. VI, 117 pages. 1994.

Vol. 1570: R. deLaubenfels, Existence Families, Functional Calculi and Evolution Equations. XV, 234 pages. 1994.

Vol. 1571: S. Yu. Pilyugin, The Space of Dynamical Systems with the C^0-Topology. X, 188 pages. 1994.

Vol. 1572: L. Göttsche, Hilbert Schemes of Zero-Dimensional Subschemes of Smooth Varieties. IX, 196 pages. 1994.

Vol. 1573: V. P. Havin, N. K. Nikolski (Eds.), Linear and Complex Analysis – Problem Book 3 – Part I. XXII, 489 pages. 1994.

Vol. 1574: V. P. Havin, N. K. Nikolski (Eds.), Linear and Complex Analysis – Problem Book 3 – Part II. XXII, 507 pages. 1994.

Vol. 1575: M. Mitrea, Clifford Wavelets, Singular Integrals, and Hardy Spaces. XI, 116 pages. 1994.

Vol. 1576: K. Kitahara, Spaces of Approximating Functions with Haar-Like Conditions. X, 110 pages. 1994.

Vol. 1577: N. Obata, White Noise Calculus and Fock Space. X, 183 pages. 1994.

Vol. 1578: J. Bernstein, V. Lunts, Equivariant Sheaves and Functors. V, 139 pages. 1994.

Vol. 1579: N. Kazamaki, Continuous Exponential Martingales and BMO. VII, 91 pages. 1994.

Vol. 1580: M. Milman, Extrapolation and Optimal Decompositions with Applications to Analysis. XI, 161 pages. 1994.

Vol. 1581: D. Bakry, R. D. Gill, S. A. Molchanov, Lectures on Probability Theory. Editor: P. Bernard. VIII, 420 pages. 1994.

Vol. 1582: W. Balser, From Divergent Power Series to Analytic Functions. X, 108 pages. 1994.

Vol. 1583: J. Azéma, P. A. Meyer, M. Yor (Eds.), Séminaire de Probabilités XXVIII. VI, 334 pages. 1994.

Vol. 1584: M. Brokate, N. Kenmochi, I. Müller, J. F. Rodriguez, C. Verdi, Phase Transitions and Hysteresis. Montecatini Terme, 1993. Editor: A. Visintin. VII. 291 pages. 1994.

Vol. 1585: G. Frey (Ed.), On Artin's Conjecture for Odd 2-dimensional Representations. VIII, 148 pages. 1994.

Vol. 1586: R. Nillsen, Difference Spaces and Invariant Linear Forms. XII, 186 pages. 1994.

Vol. 1587: N. Xi, Representations of Affine Hecke Algebras. VIII, 137 pages. 1994.

Vol. 1588: C. Scheiderer, Real and Étale Cohomology. XXIV, 273 pages. 1994.

Vol. 1589: J. Bellissard, M. Degli Esposti, G. Forni, S. Graffi, S. Isola, J. N. Mather, Transition to Chaos in Classical and Quantum Mechanics. Montecatini, 1991. Editor: S. Graffi. VII, 192 pages. 1994.

Vol. 1590: P. M. Soardi, Potential Theory on Infinite Networks. VIII, 187 pages. 1994.

Vol. 1591: M. Abate, G. Patrizio, Finsler Metrics – A Global Approach. IX, 180 pages. 1994.

Vol. 1592: K. W. Breitung, Asymptotic Approximations for Probability Integrals. IX, 146 pages. 1994.

Vol. 1593: J. Jorgenson & S. Lang, D. Goldfeld, Explicit Formulas for Regularized Products and Series. VIII, 154 pages. 1994.